ADAM KUCHARSKI is an associate professor at the London School of Hygiene & Tropical Medicine. A mathematician by training, his work on global outbreaks has included Ebola, Zika and COVID-19, and he has produced real-time analysis for multiple governments and health agencies. He is a TED senior fellow and winner of the 2016 Rosalind Franklin Award Lecture and the 2012 Wellcome Trust Science Writing Prize. The author of *The Perfect Bet*, his writing has appeared in the *Observer*, *Financial Times*, *Wired* and *New Statesman*.

'This is a hell of a moment for a book like this to come out … the principles of contagion, which, Kucharski argues, can be applied to everything from folk stories and financial crises to itching and loneliness, are suddenly of pressing interest to all of us.' *Sunday Times*

'Adam Kucharski [is] fast becoming a key voice of reason in the media circus surrounding the virus … Here he gives a clear, calm, historical overview of the mathematical ideas at the forefront of our pandemic response, where they came from and how well they stand up when you put them to the test.' Hannah Fry

'This book charts the history of this now-pivotal science, from its origins in understanding the spread of malaria at the turn of the twentieth century, to its central role in predicting the dissemination of everything from diseases to fake news in the twenty-first.' *Economist*

'Adam Kucharski's *The Rules of Contagion* is the book you want to reach for … interesting and topical.' *Guardian*

'An astonishingly bold survey of the epidemiology of more or less everything in our inter-connected world … Kucharski has pull... ...ghtest light on t... Windsor and Maidenhead ...enemy.' *Dai...*

'Lively, intriguing and elegant' *Spectator*

'*The Rules of Contagion* is a timely reminder of the importance of disease modelling. Without such models, we would be in far greater trouble battling COVID-19.' *Lancet*

'Rich in stories, *The Rules of Contagion* is a down-to-earth account of how mathematical approaches can help us better understand and, in turn, better respond to contagion in all its dynamic forms. Tackling issues from pandemics and gun violence, to financial crises and misinformation, Adam Kucharski inspires us all to think like mathematicians.' Peter Piot, Director of the London School of Hygiene & Tropical Medicine

'Fascinating exploration of the mathematics of things that go viral – not least of them viruses … Utterly timely and readable.' *Kirkus*

WELLCOME COLLECTION is a free museum and library that aims to challenge how we think and feel about health. Inspired by the medical objects and curiosities collected by Henry Wellcome, it connects science, medicine, life and art. Wellcome Collection exhibitions, events and books explore a diverse range of subjects, including consciousness, forensic medicine, emotions, sexology, identity and death.

Wellcome Collection is part of Wellcome, a global charitable foundation that exists to improve health for everyone by helping great ideas to thrive, funding over 14,000 researchers and projects in more than 70 countries.

wellcomecollection.org

The Rules of Contagion

Why Things Spread – and Why They Stop

ADAM KUCHARSKI

For Emily

This paperback edition published in 2021

First published in Great Britain in 2020 by
Profile Books Ltd
29 Cloth Fair
London
ECIA 7JQ

www.profilebooks.com

Published in association with Wellcome Collection

183 Euston Road
London NW1 2BE
www.wellcomecollection.org

1 3 5 7 9 10 8 6 4 2

Typeset in Dante by MacGuru Ltd
Printed and bound in Great Britain by
CPI Group (UK) Ltd, Croydon CR0 4YY

A CIP catalogue record for this book is available from the British Library.

ISBN 978 1 78816 020 9
eISBN 978 1 78283 430 4

Contents

Introduction

THE DAY THIS BOOK was first published in the UK, my mind was elsewhere. After seeing the morning COVID-19 data reports, I was busy trying to work out whether we'd made a major error in our analysis of the outbreak. It was 13 February 2020, and China had just reported over 15,000 new cases of the disease, a 750 per cent increase on the previous day.[1] A week earlier, our research group had released some analysis using datasets on infections within China as well as among international travellers.[2] The results suggested that control measures introduced in Wuhan in late January had led to a decline in transmission, and that the outbreak was about to peak in the city. Several media outlets had picked up on this preliminary work,[3] and for a few days, it was looking like we had made the correct call. After rising for weeks, cases finally seemed to be declining.

Then came the spike on 13 February. After almost a month of trying to extract signals from patchy available data sources, working early mornings through into the night, had we missed something crucial? It turned out that apparent surge in China was the result of health authorities changing how cases were defined, to include people with less severe symptoms. Revisiting the data, we decided that on balance, there was still enough evidence to suggest an overall decline in transmission. Still, not everyone agreed. One team in Japan reckoned the epidemic in China would peak sometime between late March and late May, with up to 2.3 million new cases diagnosed in a single day.[4]

In hindsight, the decline in COVID-19 cases in Wuhan looks obvious, just as the later declines in other cities would do. But it was anything but obvious at the time, with researchers around the world working to understand the early – and often contradictory – patterns emerging from Asia. From mid-January onwards, our research group had regular discussions with scientists and health agencies across the region, from mainland China and Japan to Hong Kong and Singapore. We swapped notes on what we knew and what we didn't, with the latter almost always outweighing the former.

One of the things that did stand out was how difficult this new virus would be to control. At the start of February, I saw some preliminary data that suggested many people with COVID-19 could spread infection before they showed clear symptoms. This meant that by the time a person became ill and got tested, they may have already infected others. These others would in turn become infectious, continuing the cycle of unseen contagion. It was exactly the sort of feature we didn't want to see in a new virus, a stealth-like ability that could easily turn a handful of cases into a large outbreak. I was visiting Melbourne for work that week and remember going for a walk through the bustling city centre, trying to picture the impact of the virus, the streets around me emptying like the images coming out of Wuhan. I'd started 2020 on honeymoon in the Galápagos Islands, where signs everywhere warned people to keep a distance of two metres from the animals. Over the following weeks, I'd watch my holiday quirk become a familiar global reality.

As well as working out how easily COVID-19 spread, there was also the question of how severe it was. By 11 February, there had been almost 45,000 confirmed cases in China, with just over 1,000 deaths reported.[5] At first glance, this might suggest around

1 in 45 cases were fatal but there were two problems. First, it took time for patients to succumb to the disease. If a hundred people with COVID-19 show up to hospital on a given day and all of them are currently alive, it doesn't mean the disease has zero fatality risk, because we need to wait and see what will eventually happen to these patients. The earliest Chinese case data, for which this 'wait-and-see' time had passed, implied that 15 per cent of cases would eventually be fatal. Which brings us to the second problem: not all cases were being detected in China. Bringing together data from international outbreaks – such as the one on the Diamond Princess cruise ship in Japan, which involved extensive testing – we estimated around 0.5 per cent of infections were fatal in China, with a much higher risk in the oldest age groups.[6]

If only a fraction of infections were fatal, and it took time for people to become severely ill, sudden reports of new COVID-19 deaths could be a sign of a much larger undetected outbreak. On 19 February, such a signal appeared when Iran reported two COVID-19 deaths, which also happened to be the first known cases in the country. Two days later, Italy reported a local outbreak in the northern Lombardy region. Many of the cases were already severely ill, indicating another big underlying outbreak.

Patchy data and undetected cases would be a recurrent problem. On 27 February, Spain reported its first local outbreak; within two weeks, some of Madrid's hospitals would be overwhelmed.[7] Meanwhile in the UK, there were reports on 10 March that a member of Parliament had been infected, with fifty-four new cases reported overall that day. My colleagues later estimated that, in reality, there were probably over five thousand new infections.[8] Across Europe, outbreaks were quietly taking hold at events and gatherings, in ski resorts and offices, homes and hospitals.[9]

Every year in late February and March, I teach on an MSc course about the spread and control of infectious diseases. As part of the course assessment, students have to run a three-day outbreak investigation. They are told that some people have fallen ill and have to piece together fragments of information – from symptoms to social contacts – to discern what has happened. While the students were analysing this fictional outbreak, our team were working with health agencies, governments and global charities, trying to do the same thing for COVID-19. What did we know about the infection? What were the benefits and downsides of different control measures? Where were the gaps in our knowledge?

Among all the uncertainties, it was clear that life would look different for a considerable period of time. That analysis we had done of the early Wuhan outbreak suggested that only 5 per cent or so of the population had been infected by the end of January.[10] If all control measures were lifted, and contagion was once again allowed to spread freely, there would still be plenty of susceptible people in the city. On 17 March, I gave a talk at a global health event, which had been hastily moved online as UK control measures came in.[11] I outlined two scenarios I thought could be plausible for the future of COVID-19. Scenario A was the depressing one: in the absence of an effective vaccine or treatment, countries might have to rely on sporadic shutdown-type measures to keep epidemics from overwhelming their health systems. Scenario B was more hopeful: some countries might be able to scale up targeted testing and combine with strict isolation and infection control, to keep outbreaks subdued – and hence away from the people most at risk – without too much disruption to the rest of society. Ultimately 2020 would be shaped by what governments chose to impose on their populations. Would they use electronic surveillance to identify infected people and make sure they stayed in isolation, like in South Korea and Taiwan? Or roll out border closures like Vietnam and New Zealand? Or try and rely on lighter-touch measures such as remote working and limits on gathering size, as in Sweden?

For me, one of the most surprising aspects of COVID-19 would be the sheer diversity in global responses. In effect, the virus had asked each country to decide what they wanted their societies to look like over the coming year – or more – and the answers varied widely. Individualist versus collectivist policies. Voluntary versus mandatory measures. Data-intensive versus privacy-protecting surveillance. Sporadic versus sustained disruption.

The pandemic would force difficult, divisive choices that would echo through the fabric of societies. Indeed, the eventual impact of COVID-19 would reach far beyond the disease itself. Alongside the coronavirus, several other forms of contagion would spread through populations during 2020. Misinformation would undermine health advice and fuel political polarisation. Economic troubles and social unrest would emerge from the disruption caused by COVID-19 control measures. Newly remote workers would fall victim to cyber-attacks and malware.[12] In the middle of this disruption, though, there was also the occasional spread of optimism. Vaccine discoveries. Treatment innovations. Knowledge. Hope.

When we think of contagion, we tend to think about things such as infectious diseases or viral online content. But outbreaks can come in many forms. They might involve things that bring harm – like violence, malware or financial crises – or benefits, like new technology or scientific progress. Some will start with tangible infections such as biological pathogens and computer viruses, others with abstract ideas and beliefs. Outbreaks will sometimes rise quickly; on other occasions they will take a while to grow. Some will create unexpected patterns and, as we wait to see what happens next, these patterns will fuel excitement, curiosity, or even fear. So why do outbreaks take off – and decline – in the way they do?

THREE AND A HALF YEARS into the First World War, a new threat to life appeared. While the German army was launching its Spring Offensive in France, across the Atlantic people had started dying at Camp Funston, a busy military base in Kansas. The cause was a new type of influenza virus, which had potentially jumped from animals into humans at a nearby farm. During 1918 and 1919, the infection would become a global epidemic – otherwise known as a pandemic – and would kill over fifty million people. The final death toll was twice as many as the entire First World War.[13]

Over the following century, there would be four more flu pandemics. Before COVID-19 emerged, people would sometimes ask me: what will the next pandemic look like? Unfortunately it was difficult to say, because even previous flu pandemics were all slightly different. There were different strains of the virus, and outbreaks hit some places harder than others. In fact, there's a saying in my field: 'if you've seen one pandemic, you've seen … one pandemic.'[14]

We face the same problem whether we're studying the spread of a disease, an online trend, or something else; one outbreak won't necessarily look like another. What we need is a way to separate features that are specific to a particular outbreak from the underlying principles that drive contagion. A way to look beyond simplistic explanations, and uncover what is really behind the outbreak patterns we observe.

That's the aim of this book. By exploring contagion across different areas of life, we'll find out what makes things spread and why outbreaks look like they do. Along the way, we'll see the connections that are emerging between seemingly unrelated problems: from banking crises, gun violence and fake news to disease evolution, opioid addiction and social inequality. As well as covering the ideas that can help us to tackle outbreaks, we'll look at the unusual situations that are changing how we think about patterns of infections, beliefs, and behaviour. The first edition of this book was written before the COVID-19 pandemic;

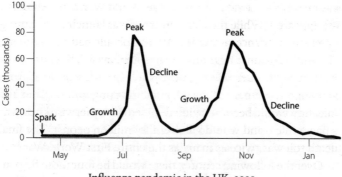

Influenza pandemic in the UK, 2009
Data from Public Health England[15]

I signed off the final page proofs at the start of December 2019, shortly before the initial cases would be reported near a seafood market in Wuhan. Although I've updated some sections to reflect the events of 2020, the central principles I cover remain the same. This is not a story of one virus, or one epidemic, but of the contagion that influences all our lives, and what we can do about it.

Let's start with the shape of an outbreak. When disease researchers hear about a new threat, one of the first things we do is draw what we call an outbreak curve – a graph showing how many cases have appeared over time. Although the shape can vary a lot, it will typically include four main stages: the spark, growth, peak, and decline. In some cases, these stages will appear multiple times; when the 'swine flu' pandemic arrived in the UK in April 2009, it grew rapidly during early summer, peaking in July, then grew and peaked again in late October (we'll find out why later in the book).

Despite the different stages of an outbreak, the focus will often fall on the spark. People want to know why it took off, how it started, and who was responsible. In hindsight, it's tempting to conjure up explanations and narratives, as if the outbreak was inevitable and could happen the same way again. But if we simply list the characteristics of successful infections or trends, we end up with an incomplete picture of how outbreaks actually work. Most things don't spark: for every influenza virus that jumps from animals to humans and spreads worldwide as a pandemic, there are millions that fail to infect any people at all. For every tweet that goes viral, there are many more that don't.

Even if an outbreak does spark, it's only the start. Try and picture the shape of a particular outbreak. It might be a disease epidemic, or the spread of a new idea. How quickly does it grow? Why does it grow that quickly? When does it peak? Is there only one peak? How long does the decline phase last?

Rather than just viewing outbreaks in terms of whether they take off or not, we need to think about how to measure them and how to predict them. Take the Ebola epidemic in West Africa back in 2014. After spreading to Sierra Leone and Liberia from Guinea, cases began to rise sharply. Our team's early analysis suggested that the epidemic was doubling every two weeks in the worst affected areas.[16] It meant that if there were currently 100 cases, there could be 200 more in a fortnight and another 400 after a month. Health agencies therefore needed to respond quickly: the longer it took them to tackle the epidemic, the larger their control efforts would need to be. In essence, opening one new treatment centre immediately was equivalent to opening four in a month's time.

Some outbreaks grow on even faster timescales. In May 2017, the WannaCry computer virus hit machines around the world, including crucial NHS systems. In its early stages, the attack was doubling in size almost every hour, eventually affecting more than 200,000 computers in 150 countries.[17] Other types of technology have taken much longer to spread. When VCRs became popular in the early 1980s, the number of owners was doubling only every 480 days or so.[18]

As well as speed, there's also the question of size: contagion

that spreads quickly won't necessarily cause a larger overall outbreak. So what causes an outbreak to peak? And what happens
after the peak? It's an issue that's relevant to many industries,
from finance and politics to technology and health. However,
not everyone has the same attitude to outbreaks. My wife
works in advertising; while my research aims to stop disease
transmission, she wants ideas and messages to spread. Although
these outlooks seem very different, it's increasingly possible to
measure and compare contagion across industries, using ideas
from one area of life to help us understand another. Over the
coming chapters, we will see why financial crises are similar to
sexually transmitted infections, why disease researchers found
it so easy to predict games like the ice bucket challenge, and
how ideas used to eradicate smallpox are helping to stop gun
violence. We will also look at the techniques we can use to slow
down transmission or – in the case of marketing – keep it going.

Our understanding of contagion has advanced dramatically
in recent years, and not just in my field of disease research. With
detailed data on social interactions, researchers are discovering
how information can evolve to become more persuasive and
shareable, why some outbreaks keep peaking – like the 2009
flu pandemic did – and how 'small-world' connections between
distant friends can help certain ideas spread widely (and yet
hinder others). At the same time, we're learning more about
how rumours emerge and spread, why some outbreaks are
harder to explain than others, and how online algorithms are
influencing our lives and infringing on our privacy.

As a result, ideas from outbreak science are now helping
to tackle threats in other fields. Central banks are using these
methods to prevent future financial crises, while technology
firms are building new defences against harmful software. In
the process, researchers are challenging long-held ideas about
how outbreaks work. When it comes to contagion, history has
shown that ideas about how things spread don't always match

reality. Medieval communities, for example, blamed the sporadic nature of outbreaks on astrological influences; influenza means 'influence' in Italian.[19]

Popular explanations for outbreaks continue to be overturned by scientific discoveries. This research is unravelling the mysteries of contagion, showing us how to avoid simplistic anecdotes and ineffective solutions. But despite this progress, coverage of outbreaks still tends to be vague: we simply hear that something is contagious or that it's gone viral. We rarely learn why it grew so quickly (or slowly), what made it peak, or what we should expect next time. Whether we're interested in spreading ideas and innovations, or stopping viruses and violence, we need to identify what's really driving contagion. And sometimes, that means rethinking everything we thought we knew about an infection.

1

A theory of happenings

WHEN I WAS THREE YEARS OLD, I lost the ability to walk. It happened gradually at first: a struggle to stand up here, a lack of balance there. But things soon deteriorated. Short distances became tricky, while slopes and stairs were near impossible. One Friday afternoon in April 1990, my parents took me and my failing legs to the Royal United Hospital in Bath. By the next morning I was seeing a neurological specialist. The initial suspect was a spinal tumour. Several days of tests followed; there were X-rays, blood samples, nerve stimulation, and a lumbar puncture to extract spinal fluid. As the results came in, the diagnosis shifted towards a rare condition known as Guillain-Barré syndrome (GBS). Named after French neurologists Georges Guillain and Jean Alexandre Barré, GBS is the result of a malfunctioning immune system. Rather than protecting my body, it had started attacking nerves, spreading paralysis.

Sometimes the sum of human wisdom is to be found, as writer Alexandre Dumas put it, within the words 'wait and hope'.[1] And that was to be my treatment, to wait and to hope. My parents were given a multicoloured party horn to check the strength of my breathing (there was no home equipment small enough for a toddler). If the horn failed to unroll when I blew, it meant the paralysis had reached the muscles that pumped air into my lungs.

There is a photo of me sitting on my grandfather's lap around this time. He is in a wheelchair. He'd caught polio in India aged twenty-five, and had been unable to walk since. I'd only ever known him like that, his strong arms wheeling uncooperative legs. In a way, it brought familiarity to this unfamiliar situation. Yet what linked us was also what separated us. We shared a symptom, but the mark of his polio was permanent; GBS, for all its misery, was usually a temporary condition.

So we waited and we hoped. The party horn never failed to unroll, and a lengthy recovery began. My parents told me GBS stood for 'Getting Better Slowly'. It was twelve months before I could walk, and another twelve before I could manage anything resembling a run. My balance would suffer for years to come.

As my symptoms faded, so did my memories. Events became distant, left behind to another life. I can no longer remember my parents giving me chocolate buttons before the needles. Or how I subsequently refused to eat them – even on a normal day – fearing what would come next. The memories of games of tag at primary school have faded too, with me spending all of lunchtime as 'it', my legs still too weak to catch the others. For the twenty-five years that followed my illness, I never really spoke about GBS. I left school, went to university, completed a PhD. GBS seemed too rare, too meaningless to bring up. Guillain-what? Barré who? The story, which I never told anyway, was over for me.

Except it wasn't quite. In 2015, I was in the Fijian capital Suva when I encountered GBS again, this time professionally. I'd been in the city to help investigate a recent dengue fever epidemic.[2] Transmitted by mosquitoes, the dengue virus causes sporadic outbreaks on islands like Fiji. Although symptoms are often mild, dengue can come with a severe fever, potentially leading to hospitalisation. During the first few months of 2014, over 25,000 people showed up at health centres in Fiji with a suspected dengue infection, putting a huge burden on the health system.

If you're imagining an office perched on a sunny beach, you're not picturing Suva. Unlike Fiji's resort-laden Western division, the capital is a port city in the southeast of the main island, Viti Levu. The two main roads of the city loop down into a peninsula, forming the horseshoe shape of a magnet, with the area in the middle attracting plenty of rain. Locals who were familiar with British weather told me that I'd feel right at home.

Another, much older, reminder of home was to follow soon after. During an introductory meeting, a colleague at the World Health Organization (WHO) mentioned that clusters of GBS had been appearing on Pacific Islands. Unusual clusters. The annual par for the disease was 1 or 2 cases per 100,000 people, but in some places they'd seen double figures.[3]

Nobody ever worked out why I got GBS. Sometimes it follows an infection – GBS has been linked to flu and pneumonia, as well as other diseases[4] – but sometimes there's no clear trigger. In my case, the syndrome was just noise, a random blip in the grand scheme of human health. But in the Pacific during 2014/15, GBS represented a signal, just like birth defects would soon do in Latin America.

Behind these new signals lay the Zika virus, named after the Zika Forest in southern Uganda. A close relative of the dengue virus, Zika was first identified in the forest's mosquitoes in 1947. In the local language, Zika means 'overgrown'[5] and grow it would, from Uganda to Tahiti to Rio de Janeiro and beyond. Those signals in the Pacific and Latin America in 2014 and 2015 would gradually become clearer. Researchers found increasing evidence of a link between Zika infection and neurological conditions: as well as GBS, Zika seemed to lead to pregnancy complications. The main concern was microcephaly, where babies develop a smaller brain than usual, resulting in a smaller skull.[6] This can cause a host of serious health issues, including seizures and intellectual disabilities.

In February 2016, triggered by the possibility that Zika was causing microcephaly,[7] WHO announced that the infection was a Public Health Emergency of International Concern, or PHEIC (pronounced 'fake'). Early studies had suggested that for every 100 Zika infections during pregnancy, there could be between 1 and 20 babies with microcephaly.[8] Although microcephaly would become the primary concern about Zika, it was GBS that first brought the infection into health agencies' focus, as well as into mine. Sitting in my temporary office in Suva in 2015, I realised that this syndrome, which had shaped so much of my childhood, was one I knew almost nothing about. My ignorance was mostly self-inflicted, with some (entirely understandable) assistance from my parents: it was years before they told me GBS could be fatal.

At the same time, the health world was facing a much deeper ignorance. Zika was generating a huge volume of questions, few of which could yet be answered. 'Rarely have scientists engaged with a new research agenda with such a sense of urgency and from such a small knowledge base,' wrote epidemiologist Laura Rodrigues in early 2016.[9] For me, the first challenge was to understand the dynamics of these Zika outbreaks. How easily did the infection spread? Were the outbreaks similar to dengue ones? How many cases should we expect?

To answer these questions, our research group started to develop mathematical models of the outbreaks. Such approaches are now commonly used in public health, as well as appearing in several other fields of research. But where do these models originally come from? And how do they actually work? It's a story that starts in 1883 with a young army surgeon, a water tank and an angry staff officer.

RONALD ROSS HAD WANTED to be a writer, but his father pushed him into medical school. His studies at St Bartholomew's in London struggled to compete with his poems, plays and music, and when Ross took his two qualifying exams in 1879, he passed only the surgery one. This meant he could not join the colonial Indian Medical Service, his father's preferred career path.[10]

Unable to practice general medicine, Ross spent the next year sailing the Atlantic as a ship's surgeon. Eventually he passed his remaining medical exam and scraped into the Indian Medical Service in 1881. After two years in Madras, Ross moved to Bangalore to take up a post as Garrison Surgeon in September 1883. From his comfortable colonial viewpoint, he claimed it was a 'picture of pleasure', a city of sun, gardens and pillared villas. The only problem, as he saw it, was the mosquitoes. His new bungalow seemed to attract far more than the other army rooms. He suspected it was something to do with the water barrel sitting outside his window, which was surrounded by the insects.

Ross's solution was to tip over the tank, destroying the mosquitoes' breeding ground. It seemed to work: without the stagnant water, the insects left him alone. Spurred on by his successful experiment, he asked his staff officer if they could remove the other water tanks too. And while they were at it, why not also get rid of the vases and tins that lay scattered around the mess? If the mosquitoes had nowhere to breed, they would have little option but to move on. The officer wasn't interested. 'He was very scornful and refused to allow men to deal with them,' Ross later wrote, 'for he said it would be upsetting to the order of nature, and as mosquitoes were created for some purpose it was our duty to bear with them.'

The experiment would turn out to be the first in a lifelong analysis of mosquitoes. The second study would come over a decade later, inspired by a conversation in London. In 1894, Ross had travelled back to England for a one-year sabbatical. The city had changed a lot since his last visit: Tower Bridge had been completed, Prime Minister William Gladstone had just resigned, and the country was about to get its first film parlour.[11] When Ross arrived, though, his mind was focused elsewhere. He wanted to catch up on the latest malaria research. In India, people regularly fell ill with the disease, which could lead to fever, vomiting, and sometimes death.

Malaria is one of the oldest diseases known to humanity. In fact, it may have been with us for our entire history as a species.[12] However, its name comes from Medieval Italy. Those who caught a fever would often blame 'mala aria': bad air.[13] The name stuck, as did the blame. Although the disease was eventually traced to a parasite called *Plasmodium*, when Ross arrived back in England the cause of its spread was still a mystery.

In London, Ross called on biologist Alfredo Kanthack at St Bartholomew's, hoping to learn about developments he may have missed while in India. Kanthack said that if Ross wanted to know more about parasites like malaria, he should go and speak to a doctor called Patrick Manson. For several years, Manson had researched parasites in southeastern China. While there, he had discovered how people get infected with a particularly nasty family of microscopic worms called *filariae*. These parasites were small enough to get into a person's bloodstream and infect their lymph nodes, causing fluid to accumulate within the body. In severe cases, a person's limbs could swell to many times their natural size, a condition known as elephantiasis. As well as identifying how the *filariae* caused disease, Manson had shown that when mosquitoes fed on infected humans, they could also suck up the worms.[14]

Manson invited Ross into his lab, teaching him how to find parasites like malaria in infected patients. He also pointed Ross to recent academic papers he'd missed while out in India. 'I visited him often and learnt all he had to tell me,' Ross later recalled. One winter afternoon, they were walking down Oxford Street, when Manson made a comment that would transform Ross's career. 'Do you know,' he said, 'I have formed the theory that mosquitoes carry malaria just as they carry *filariae*.'

Other cultures had long speculated about a potential link between mosquitoes and malaria. British geographer Richard Burton noted that in Somalia, it was often said that mosquito bites brought on deadly fevers, though Burton himself dismissed the idea. 'The superstition probably arises from the fact that mosquitoes and fevers become formidable about the same time,' he wrote in 1856.[15] Some people had even developed treatments for malaria, despite not knowing what caused the disease. In the fourth century, Chinese scholar Ge Hong described how the qinghao plant could reduce fevers. Extracts of this plant now form the basis for modern malaria treatments.[16] (Other attempts were less successful: the word 'abracadabra' originated as a Roman spell to ward off the disease.[17])

Ross had heard the speculation linking mosquitoes and malaria, but Manson's argument was the first to really convince him. Just as mosquitoes ingested those tiny worms when they fed on human blood, Manson reckoned that they could also pick up malaria parasites. These parasites then reproduced within the mosquito before somehow making their way back into humans. Manson suggested that drinking water might be the source of infection. When Ross returned to India, he set out to test the idea, with an experiment that would be unlikely to pass a modern ethics board.[18] He got mosquitoes to feed on an infected patient then lay eggs in a bottle of water; once the eggs had hatched, he paid three people to drink the water. To his disappointment, none of them got malaria. So how did the parasites get into people?

Ross eventually wrote to Manson with a new theory, suggesting that the infection might spread through mosquito bites. The mosquitoes injected some saliva with each bite: maybe this was enough to let the parasites in? Unable to recruit enough human volunteers for another study, Ross experimented with birds. First, he collected some mosquitoes and got them to feed on the blood of an infected bird. Then he let these mosquitoes bite healthy birds, which soon came down with the disease as well. Finally, he dissected the saliva glands of the infected mosquitoes, where he found malaria parasites. Having discovered the true route of transmission, he realised just how absurd their previous theories had been. 'Men and birds don't go about eating dead mosquitoes,' he told Manson.

In 1902, Ross received the second ever Nobel Prize for medicine for his work on malaria. Despite contributing to the discovery, Manson did not share the award. He only found out that Ross had won when he saw it in a newspaper.[19] The once close friendship between mentor and student gradually splintered into a sharp animosity. Though he was a brilliant scientist, Ross could be a divisive colleague. He got into a series of disputes with his rivals, often involving legal action. In 1912, he even threatened to sue Manson for libel.[20] The offence? Manson had written a complimentary reference letter for another researcher, who was taking up a professorship that Ross had recently vacated. Manson did not rise to the argument, choosing to apologise instead. 'It takes two fools to make a quarrel,' as he later put it.[21]

Ross would continue to work on malaria without Manson. In the process, he'd find a new outlet for his single-minded stubbornness, and a new set of opponents. Having discovered how malaria spread, he wanted to demonstrate that it could be stopped.

MALARIA ONCE HAD A MUCH BROADER reach than it does today. For centuries, the disease stretched across Europe and North America, from Oslo to Ontario. Even as temperatures dropped during the so-called Little Ice Age in the seventeenth and eighteenth centuries, the biting cold of winter would still be followed by the biting mosquitoes of summer.[22] Malaria was endemic in many temperate countries, with ongoing transmission and a regular stream of new cases from one year to the next. Eight of Shakespeare's plays include mentions of 'ague', a medieval term for malarial fever. The salt marshes of Essex, northeast of London, had been a notorious source of disease for centuries; when Ronald Ross was a student, he'd treated a woman who picked up malaria there.

Having made the link between insects and infections, Ross argued that removing mosquitoes was the key to controlling malaria. His experiences in India – like the experiment with the water tank in Bangalore – had persuaded him that mosquito numbers could be reduced. But the idea went against popular wisdom. It was impossible to get rid of every last mosquito, went the argument, which meant there would always be some insects left, and hence potential for malaria to spread. Ross acknowledged that some mosquitoes would remain, but he believed that malaria transmission could still be stopped. From Freetown to Calcutta, his suggestions were at best ignored and at worst derided. 'Everywhere, my proposal to reduce mosquitoes in towns was treated only with ridicule,' he later recalled.

In 1901, Ross had led a team to Sierra Leone to try and put his mosquito control ideas into practice. They cleared away cartloads of tins and bottles. They poisoned the standing water mosquitoes loved to breed in. And they filled potholes so 'death-dealing street-puddles', as Ross called them, couldn't form on the roads. The results were promising: when Ross visited again a year later, there were far fewer mosquitoes. However, he had warned health authorities the effect would only last if the control measures continued. Funding for the clean up had come from a wealthy Glaswegian donor. When the money ran out, enthusiasm waned, and mosquito numbers increased once again.

Ross had more success advising the Suez Canal Company the following year. They'd been seeing around 2,000 malaria cases a year in the Egyptian city of Ismailia. After intensive mosquito reduction efforts, this number fell below a hundred. Mosquito control was also proving effective elsewhere. When the French had attempted to build a canal in Panama during the 1880s, thousands of workers had died from malaria, as well as yellow fever, another mosquito-borne infection. In 1905, with the Americans now leading the Panama project, US Army Colonel William Gorgas oversaw an intensive mosquito control campaign, making it possible to complete the canal.[23] Meanwhile further south, physicians Oswaldo Cruz and Carlos Chagas were spearheading anti-malaria programmes in Brazil, helping to reduce cases among construction workers.[24]

Despite these projects, many remained sceptical about mosquito control. Ross would need a stronger argument to persuade his peers. To make his point, he would eventually turn to mathematics. During those early years in the Indian Medical Service, he'd taught himself the subject to a fairly advanced level. The artist in him admired its elegance. 'A proved proposition was like a perfectly balanced picture,' he later suggested. 'An infinite series died away into the future like the long-drawn variations of a sonata.' Realising how much he liked the subject, he regretted not studying it properly at school. He was now too far into his career to change direction; what use was mathematics to someone working in medicine? 'It was the unfortunate passion of a married man for some beautiful but inaccessible lady,' as he put it.

Ross put the intellectual affair behind him for a while, but returned to the subject after his mosquito discovery. This time, he found a way to make his mathematical hobby useful to his professional work. There was a vital question he needed to answer: was it really possible to control malaria without removing every mosquito? To find out, he developed a simple conceptual model of malaria transmission. He started by calculating how many new human malaria infections there might be each month, on average, in a given geographic area. This meant breaking down the process of transmission into its basic components. For transmission to occur, he reasoned, there first needs to be at least one human in the area who is infectious with malaria. As an example, he picked a scenario where there was one infectious person in a village of 1,000. For the infection to pass to another human, an *Anopheles* mosquito would have to bite this infectious human. Ross reckoned only 1 in 4 mosquitoes would manage to bite someone. So if there were 48,000 mosquitoes in an area, he'd expect only 12,000 to bite a person. And because only 1 person in 1,000 was initially infectious, on average only 12 of those 12,000 mosquitoes would bite that one infectious person and pick up the parasite.

It takes some time for the malaria parasite to reproduce within a mosquito, so these insects would also have to survive long enough to become infectious. Ross assumed only 1 in every 3 mosquitoes would make it this far, which meant that of the 12 mosquitoes with the parasite, only 4 would eventually become infectious. Finally, these mosquitoes would need to bite another human to pass on the infection. If, again, only 1 in 4 of them successfully fed off a human, this would leave a single infectious mosquito to transmit the virus. Ross's calculation showed that even if there were 48,000 mosquitoes in the area, on average they would generate only one new human infection.

If there were more mosquitoes, or more infected humans, by the above logic we'd expect more new infections per month. However, there is a second process that counteracts this effect: Ross estimated that around 20 per cent of humans infected with malaria would recover each month. For malaria to remain endemic in the population, these two processes – infection and recovery – would need to balance each other out. If the recoveries outpaced the rate of new infections, the level of disease eventually would decline to zero.

This was his crucial insight. It wasn't necessary to get rid of every last mosquito to control malaria: there was a critical

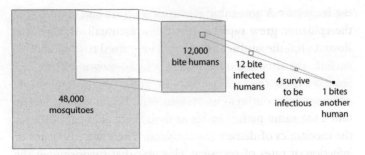

Ross calculated that even if there were 48,000 mosquitoes
in a village that contained someone infected with malaria,
it might only result in one additional human case

mosquito density, and once the mosquito population fell below
this level, the disease would fade away by itself. As Ross put it,
'malaria cannot persist in a community unless the *Anophelines*
are so numerous that the number of new infections compen-
sates for the number of recoveries.'

When he wrote up the analysis in his 1910 book *The Prevention of Malaria*, Ross acknowledged that his readers might not
follow all of his calculations. Still, he believed that they would
be able to appreciate the implications. 'The reader should make
a careful study of those ideas,' he wrote, 'and will, I think, have
little difficulty in understanding them, though he may have
forgotten most of his mathematics'. Keeping with the mathematical theme, he called his discovery the 'mosquito theorem'.

The analysis showed how malaria could be controlled, but
it also included a much deeper insight, which would revolu-
tionise how we look at contagion. As Ross saw it, there were
two ways to approach disease analysis. Let's call them 'descriptive' and 'mechanistic' methods. In Ross's era, most studies
used descriptive reasoning. This involved starting with real-life
data and working backwards to identify predictable patterns.
Take William Farr's analysis of a London smallpox outbreak in

the late 1830s. A government statistician, Farr had noticed that the epidemic grew rapidly at first, but eventually this growth slowed until the outbreak peaked, then started to decline. This decline was almost a mirror image of the growth phase. Farr plotted a curve through case data to capture the general shape; when another outbreak started in 1840, he found it followed much the same path.[25] In his analysis, Farr didn't account for the mechanics of disease transmission. There were no rates of infection or rates of recovery. This isn't that surprising: at the time nobody knew that smallpox was a virus. Farr's method therefore focused on what shape epidemics take, not why they take that shape.[26]

In contrast, Ross adopted a mechanistic approach. Rather than taking data and finding patterns that could describe the observed trends, he started by outlining the main processes that influenced transmission. Using his knowledge of malaria, he specified how people became infected, how they infected others, and how quickly they recovered. He summarised this conceptual model of transmission using mathematical equations, which he then analysed to make conclusions about likely outbreak patterns.

Because his analysis included specific assumptions about the transmission process, Ross could tweak these assumptions to see what might happen if the situation changed. What effect might mosquito reduction have? How quickly would the disease disappear if transmission declined? Ross's approach meant he could look forward and ask 'what if?', rather than just searching for patterns in existing data. Although other researchers had made rough attempts at this type of analysis before, Ross brought the ideas together into a clear, comprehensive theory.[27] He showed how to examine epidemics in a dynamic way, treating them as a series of interacting processes rather than a set of static patterns.

Descriptive and mechanistic methods – one looking back

and the other forward – should in theory converge to the same answer. Take the descriptive approach. With enough real-life data, it would be possible to estimate the effect of mosquito control: tip over a water tank, or remove mosquitoes in some other way, and we can observe what happens. Conversely, the predicted effect of mosquito control in Ross's mathematical analysis should ideally match the real impact of such measures. If a control strategy genuinely works, both methods should tell us that it does. The difference is that with Ross's mechanistic approach, we don't need to knock over water tanks to estimate what effect it might have.

Mathematical models like Ross's often have a reputation for being opaque or complicated. But in essence, a model is just a simplification of the world, designed to help us understand what might happen in a given situation. Mechanistic models are particularly useful for questions that we can't answer with experiments. If a health agency wants to know how effective their disease control strategy was, they can't go back and rerun the same epidemic without it. Likewise, if we want to know what a future pandemic might look like, we can't deliberately release a new virus and see how it spreads. Models give us the ability to examine outbreaks without interfering with reality. We can explore how things like transmission and recovery affect the spread of infection. We can introduce different control measures – from mosquito removal to vaccination – and see how effective they might be in different situations.

In the early twentieth century, this approach was exactly what Ross needed. When he announced that *Anopheles* mosquitoes spread malaria, many of his peers were unconvinced that mosquito control would reduce the disease. This made descriptive analysis problematic: it's tricky to assess a control measure if it's not being used. Thanks to his new model, however, Ross had convinced himself that long-term mosquito reduction would work. The next challenge was convincing everyone else.

From a modern viewpoint, it might seem strange that there was so much opposition to Ross's ideas. Although the science of epidemiology was expanding, creating new ways to analyse disease patterns, the medical community didn't view malaria in the same way that Ross did. Fundamentally, it was a clash of philosophies. Most physicians thought about malaria in terms of descriptions: when looking at outbreaks, they dealt in classifications rather than calculus. But Ross was adamant that the processes behind disease epidemics needed to be quantified. 'Epidemiology is in fact a mathematical subject,' he wrote in 1911, 'and fewer absurd mistakes would be made regarding it (for example, those regarding malaria) if more attention were given to the mathematical study of it.'[28]

It would take many more years for mosquito control to be widely adopted. Ross would not live to see the most dramatic reductions in malaria cases: the disease remained in England until the 1950s, and was only eliminated from continental Europe in 1975.[29] Although his ideas eventually started to catch on, he lamented the delay. 'The world requires at least ten years to understand a new idea,' he once wrote, 'however important or simple it may be.'

It wasn't just Ross's practical efforts that would spread over time. One of the team on that 1901 expedition to Sierra Leone had been Anderson McKendrick, a newly qualified doctor from Glasgow. McKendrick had top-scored in the Indian Medical Service exams and was scheduled to start his new job in India after the Sierra Leone trip.[30] On the ship back to Britain, McKendrick and Ross talked at length about the mathematics of disease. The pair continued to exchange ideas over the following years. Eventually, McKendrick would pick up enough maths to try and build on Ross's analysis. 'I have read your work in your capital book,' he told Ross in August 1911. 'I am trying to reach the same conclusions from differential equations, but it is a very elusive business, and I am having to extend mathematics in new directions. I doubt whether I shall be able to get what I want, but "a man's reach must exceed his grasp".'[31]

McKendrick would develop a scathing view of statisticians like Karl Pearson, who relied heavily on descriptive analysis rather than adopting Ross's mechanistic methods. 'The Pearsonians have as usual made a frightful hash of the whole business,' he told Ross after reading a flawed analysis of malaria infections. 'I have no sympathy with them, or their methods.'[32] Traditional descriptive approaches were an important part of medicine – and still are – but they have limitations when it comes to understanding the process of transmission. McKendrick believed the future of outbreak analysis lay with a more dynamic way of thinking. Ross shared this view. 'We shall end by establishing a new science,' he once told McKendrick. 'But first let you and me unlock the door and then anybody can go in who likes.'[33]

ONE SUMMER EVENING IN 1924, William Kermack's experiment exploded, spraying corrosive alkali solution into his eyes. A chemist by training, Kermack had been investigating the methods commonly used to study spinal fluids. He was working alone in Edinburgh's Royal College of Physicians Laboratory that evening, and would eventually spend two months in hospital with his injuries. The accident left the 26-year-old Kermack completely blind.[34]

During his stay in hospital, Kermack asked friends and nurses to read mathematics to him. Knowing that he could no longer see, he wanted to practise getting information another way. He had an exceptional memory and would work through mathematical problems in his head. 'It was incredible to find how much he could do without being able to put anything down on paper,' remarked William McCrea, one of his colleagues.

After leaving hospital, Kermack continued to work in science but shifted his focus to other topics. He left his chemical experiments behind, and began to develop new projects. In particular, he started to work on mathematical questions with Anderson McKendrick, who had risen to become head of the Edinburgh laboratory. Having served in India for almost two decades, McKendrick had left the Indian Medical Service in 1920 and moved to Scotland with his family.

Together, the pair extended Ross's ideas to look at epidemics in general. They focused their attention on one of the most important questions in infectious disease research: what causes epidemics to end? The pair noted that there were two popular explanations at the time. Either transmission ceased because there were no susceptible people left to infect, or because the pathogen itself became less infectious as the epidemic progressed. It would turn out that, in most situations, neither explanation was correct.[35]

Like Ross, Kermack and McKendrick started by developing a mathematical model of disease transmission. For simplicity, they assumed the population mixed randomly in their model. Like marbles being shaken in a jar, everyone in the population has an equal chance of meeting everyone else. In the model, the epidemic sparks with a certain number of infectious people, and everyone else susceptible to infection. Once someone has recovered from infection, they are immune to the disease. We can therefore put the population into one of three groups, based on their disease status:

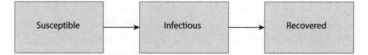

Given the names of the three groups, this is commonly known as the 'SIR model'. Say a single influenza case arrives in a population of 10,000 people. If we simulate a flu-like epidemic using the SIR model, we get the following pattern:

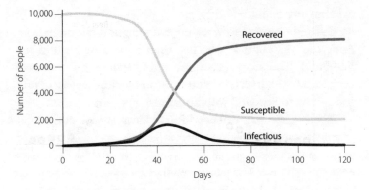

Simulated influenza outbreak using the SIR model

The simulated epidemic takes a while to grow because only one person is infectious at the start, but it still peaks within fifty days. And by eighty days, it's all but over. Notice that at the end of the epidemic, there are still some susceptible people left. If everyone had been infected, then all 10,000 people would have eventually ended up in the 'recovered' group. Kermack and McKendick's model suggests that this doesn't happen: outbreaks can end before everyone picks up the infection. 'An epidemic, in general, comes to an end before the susceptible population has been exhausted,' as they put it.

Why doesn't everyone get infected? It's because of a transition that happens mid-outbreak. In the early stages of an epidemic, there are lots of susceptible people. As a result, the number of people who become infected each day is larger than the number who recover, and the epidemic grows. Over time, however, the pool of susceptible people shrinks. When this pool gets small enough, the situation flips around: there are more recoveries than new infections each day, so the epidemic begins to decline. There are still susceptible people out there who could be infected, but there are so few left that an infectious person is more likely to recover than meet one.

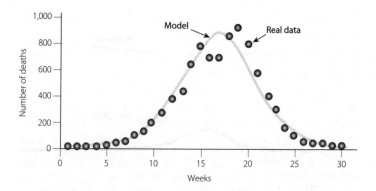

The 1906 plague outbreak in Bombay, with SIR
model shown alongside real data

To illustrate the effect, Kermack and McKendrick showed
how the SIR model could reproduce the dynamics of a 1906
outbreak of plague in Bombay (now Mumbai). In the model,
the pathogen remains equally infectious over time; it is the shifting numbers of susceptible and infectious people that lead to
the rise and fall.

The crucial change happens at the peak of the epidemic.
At this point, there are so many immune people – and so few
susceptible – that the epidemic cannot continue to grow. The
epidemic will therefore turn over and start its decline.

When there are enough immune people to prevent transmission, we say that the population has acquired 'herd immunity'.
The phrase was originally coined by statistician Major Greenwood in the early twentieth century (Major was his first name,
his army rank was actually captain).[36] Psychologists had previously used 'herd instinct' to describe groups that acted as a
collective rather than as individuals.[37] Likewise, herd immunity
meant that the population as a whole could block transmission,
even if some individuals were still susceptible.

The concept of herd immunity would find popularity several

decades later, when people realised it could be a powerful tool for disease control. During an epidemic, people naturally move out of the susceptible group as they become infected. But for many infections, health agencies can move people out of this group deliberately, by vaccinating them. Just as Ross suggested malaria could be controlled without removing every last mosquito, herd immunity makes it possible to control infections without vaccinating the entire population. There are often people who cannot be vaccinated – such as newborn babies or those with compromised immune systems – but herd immunity allows vaccinated people to protect these vulnerable unvaccinated groups as well as themselves.[38] And if diseases can be controlled through vaccination, they can potentially be eliminated from a population. This is why herd immunity has found its way into the heart of epidemic theory. 'The concept has a special aura,' as epidemiologist Paul Fine once put it.[39]

As well as looking at why epidemics end, Kermack and McKendrick were also interested in the apparently random occurrence of outbreaks. Analysing their model, they found that transmission was highly sensitive to small differences in the characteristics of the pathogen or human population. This explains why large outbreaks can seemingly appear from nowhere. According to the SIR model, outbreaks need three things to take off: a sufficiently infectious pathogen, plenty of interactions between different people, and enough of the population who are susceptible. Near the critical herd immunity threshold, a small change in one of these factors can be the difference between a handful of cases and a major epidemic.

THE FIRST REPORTED OUTBREAK of Zika began on the Micronesian island of Yap in early 2007. Before then, only fourteen human cases of Zika had ever been spotted, scattered across Uganda, Nigeria, and Senegal. But the Yap outbreak was different. It was

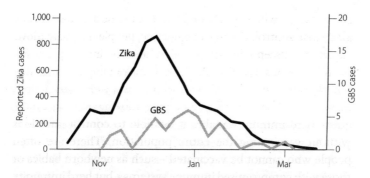

Zika and Guillain-Barré syndrome cases in French Polynesia, 2013/14
Data: French Polynesia Ministry of Health[40]

explosive, with most of the island getting infected, and completely unexpected. The little-known virus from the overgrown forest was apparently entering a new era. 'Public health officials should be aware of the risk of further expansion of Zika virus transmission,' concluded epidemiologist Mark Duffy and his colleagues in their outbreak report.[41]

In Yap, Zika had been a curiosity rather than a major threat. Despite lots of people getting a fever or rash, nobody ended up in hospital. That changed when the virus arrived on the much larger islands of French Polynesia in late 2013. During the resulting outbreak, forty-two people with Guillain-Barré Syndrome arrived at the main city hospital in Papeete, on the northern coast of Tahiti. The GBS cases cropped up slightly later than the main Zika outbreak, which is what we'd expect for a syndrome that takes a couple of weeks to appear after an infection. Speculation about a possible link was confirmed when local scientist Van-Mai Cao-Lormeau and her colleagues discovered that almost all the GBS cases had recently been infected with Zika.[42]

As in Yap, the French Polynesia outbreak had been huge, with the majority of the population infected. And like Yap, the outbreak had been very brief, with the bulk of cases appearing over a few weeks. Given that our team had spent 2014–15 developing mathematical models to analyse dengue in the Pacific, we decided to turn our attention to Zika as well. Unlike the plain-coloured *Anophelines* that can fly miles to spread malaria, dengue and Zika are both spread by *Aedes* mosquitoes, best known for being stripey and lazy ('aedes' means 'house' in Latin). As a result, the infection generally spreads when humans move from one place to another.[43]

When we tried to get our model simulations to reproduce the dynamics of Zika in French Polynesia, we realised there must have been a large, dengue-like rate of spread to generate such an explosive outbreak.[44] The short span of the outbreak stood out even more when we considered the delays involved in the infection process. During each cycle of transmission, the virus has to get from a human into a mosquito then back into another human.

While analysing transmission rates in French Polynesia, we also estimated how many people were already infected when the first cases were reported in October 2013. Our model suggested there had been several hundred infections by this point, meaning the virus probably arrived in the country weeks if not months earlier. This result would link into another mystery: how did the Zika virus reach Latin America? After the first cases were reported in Brazil during May 2015, there was a lot of speculation about when the infection had been introduced to the continent, and by whom. One early hypothesis pointed to the FIFA World Cup, held in Brazil during June/July 2014, which had attracted over three million football fans from around the globe. Another candidate was the Va'a sprint canoe championship, held in Rio de Janeiro during August 2014. Unlike the World Cup, this smaller event had included a team from French Polynesia. So which explanation was most plausible?

According to evolutionary biologist Nuno Faria and his colleagues, neither theory was particularly good.[45] Based on the genetic diversity of Zika viruses circulating in Latin America by 2016, they reckoned that the infection was introduced much earlier than previously thought. The virus probably hit the continent in mid-to-late 2013. Although too early for the canoe championship or World Cup, the time range coincided with the Confederations Cup, a regional football tournament held in June 2013. What's more, French Polynesia was one of the countries competing.

There was just one gap in the theory: the Confederations Cup occurred five months before the first Zika cases were reported in French Polynesia. But if the outbreak in French Polynesia had in reality started earlier than October 2013 – as our analysis suggested – it was just about plausible that it could have spread to Latin America during that summer. (Of course, we should be cautious about trying too hard to find a sport-shaped prologue for the Zika story: there's always a chance that it was just a random person in the Pacific taking a random flight to Brazil sometime in 2013.)

As well as analysing past outbreaks, we can use mathematical models to look at what might happen in future. This can be particularly useful for health agencies faced with difficult decisions during an outbreak. One such difficulty came in December 2015, when Zika reached the Caribbean island of Martinique. A big concern was the island's ability to handle GBS cases: if patients' lungs failed, they would need to be put on ventilators. At the time, Martinique only had eight ventilators for a population of 380,000. Would it be enough?

To find out, researchers at Institut Pasteur in Paris developed a model of Zika transmission on the island.[46] The crucial thing they wanted to know was the overall shape of the outbreak. GBS cases who required a ventilator typically stayed on it for several weeks, so a short outbreak with a large peak could overwhelm the health system, while a longer, flatter outbreak would not. At the very start of the Martinique outbreak, there hadn't been many cases, so the team used data from French Polynesia as a starting point. Of the forty-two GBS cases reported there in 2013/14, twelve had required ventilators. According to the Pasteur model, this meant they could have a big problem. If the outbreak in Martinique followed the same pattern as French Polynesia, the island would probably need nine ventilators, one more than they had available.

Fortunately, the Martinique outbreak wouldn't be the same. As new data came in, it became clear that the virus wasn't spreading as quickly as it had in French Polynesia. At the peak of the outbreak, the researchers expected there would be around three GBS cases needing ventilators. Even in the worst-case scenario, they estimated that seven ventilators would be enough. Their conclusion about this upper limit turned out to be correct: at the peak of the outbreak, there were five GBS cases on ventilators. Overall, there were thirty cases of GBS during the outbreak, with two deaths. Without adequate medical facilities, the outcome could have been much worse.[47]

These Zika studies are just a few illustrations of how Ross's methods have influenced our understanding of infectious diseases. From predicting the shape of an outbreak to evaluating control measures, mechanistic models have become a fundamental part of how we study contagion today. Researchers are using models to help health agencies respond to a whole host of outbreaks, from malaria and Zika to HIV and Ebola, in locations ranging from remote islands to conflict zones.

Ross would no doubt be glad to see how influential his ideas have been. Despite winning a Nobel Prize for his discovery that mosquitoes transmit malaria, he did not view this as his biggest achievement. 'In my own opinion my principal work has been to establish the general laws of epidemics,' he once wrote.[48] And he didn't just mean disease epidemics.

ALTHOUGH KERMACK AND MCKENDRICK would later extend Ross's mosquito theorem to other types of infections, Ross had wider ambitions. 'As infection is only one of many kinds of events which may happen to such organisms, we shall deal with "happenings" in general,' he wrote in the second edition of *The Prevention of Malaria*. Ross proposed a 'Theory of Happenings' to describe how the number of people affected by something – whether a disease or another event – might change over time.

Ross suggested that there are two main types of happening. The first type affects people independently: if it happens to you, it generally won't increase or decrease the chances of it happening to someone else afterwards. According to Ross, this could include things like non-infectious diseases, accidents or divorce.[49] For example, suppose there is a new condition that can randomly affect anyone, but at first nobody in the population has it. If each person has a certain chance of becoming affected every year – and remains affected from that point onwards – we'd expect to see a rising pattern over time.

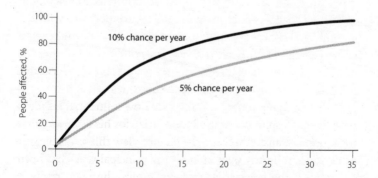

Growth of an independent happening over time. Example shows what would happen if everyone had a 5 per cent or 10 per cent chance of being affected per year

The curve gradually flattens off, though, because the size of the unaffected group shrinks over time. Each year, a proportion of people who were previously unaffected get the condition, but because there are fewer and fewer of such people over time, the overall total doesn't grow so much later on. If the chance of being affected each year is lower, the curve will grow more slowly initially, but still eventually plateau. In reality, the curve won't necessarily level off at 100 per cent: the final amount of people affected will depend on who is initially 'susceptible' to the happening.

As an illustration, consider home ownership in the UK. Of people who were born in 1960, very few were homeowners by the age of twenty, but the majority had owned a house by the time they were thirty years old. In contrast, people who were born in 1980 or 1990 had a much lower chance of becoming a homeowner during each year of their twenties. If we plot the proportion of people who become homeowners over time, we can see how quickly ownership grows in different age groups.

Of course, home ownership isn't completely random

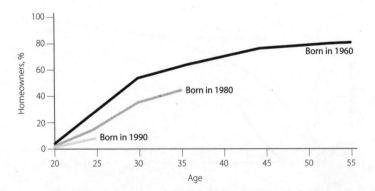

Percentage of people who were homeowners
by a given age, based on year of birth
Data: Council of Mortgage Lenders[50]

– factors such as inheritance influence people's chance of buying – but the overall pattern lines up with Ross's concept of an independent happening. On average, one twenty-year-old becoming a homeowner won't have much effect on whether another gets on the housing ladder. As long as events occur independently of one another at a fairly consistent rate, this overall pattern won't vary much. Whether we plot the amount of people who are on the housing ladder by a certain age, or the chance your bus has arrived after a certain time waiting, we'll get a similar picture.

Independent happenings are a natural starting point, but things get more interesting when events are contagious. Ross called these types of events 'dependent happenings', because what happens to one person depends on how many others are currently affected. The simplest type of outbreak is one where affected people pass the condition on to others, and once affected, people remain so. In this situation, the happening will gradually permeate through the population. Ross noted that such epidemics would follow the shape of a 'long-drawn-out

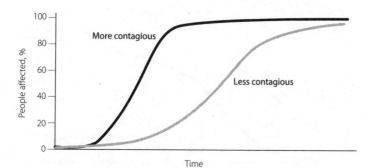

Illustrative example of the S-shaped growth of a dependent happening, based on Ross's model. The plot shows the growth of a more contagious and less contagious happening

letter S'. The number of people affected grows exponentially at first, with the number of new cases rising faster and faster over time. Eventually, this growth slows down and levels off.

The assumption that people remain affected indefinitely doesn't usually apply to infectious diseases, because people may recover, receive treatment, or die from the infection. But it can capture other kinds of spread. The S-shaped curve would later become popular in sociology, after Everett Rogers featured it in his 1962 book *Diffusion of Innovations*.[51] He noted that the initial adoption of new ideas and products generally followed this shape. In the mid twentieth century, the diffusion of products, like radios and refrigerators, all traced out an S-curve; later on, televisions, microwave ovens and mobile phones would do so as well.

According to Rogers, four different types of people are responsible for the growth of a product: initial uptake comes from 'innovators', followed by 'early adopters', then the majority of the population, and finally 'laggards'. His research into innovations mostly followed this descriptive approach, starting with the S-curve and trying to find possible explanations.

Ross had worked in the opposite direction. He'd used his

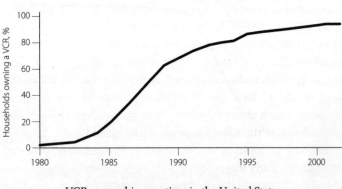

VCR ownership over time in the United States
Data: Consumer Electronics Association

mechanistic reasoning to derive the curve from scratch, showing that the spread of such happenings would inevitably lead to this pattern. Ross's model also gives us an explanation for why the adoption of new ideas gradually slows down. As more people adopt, it becomes harder and harder to meet someone who has not yet heard about the idea. Although the overall number of adopters continues to grow, there are fewer and fewer people adopting it at each point in time. The number of new adoptions therefore begins to decline.

In the 1960s, marketing researcher Frank Bass developed what was essentially an extended version of Ross's model.[52] Unlike Rogers's descriptive analysis, Bass used his model to look at the timescale of adoption as well as the overall shape. By thinking about the way people might adopt innovations, Bass was able to make predictions about the uptake of new technology. In Rogers's curve, innovators are responsible for the first 2.5 per cent of adoptions, with everyone else in the remaining 97.5 per cent. These values are somewhat arbitrary: because Rogers relied on a descriptive method, he needed to know the full shape of the S-curve; it was only possible to categorise people once an idea had been fully adopted. In contrast, Bass could use the early shape of the adoption curve to estimate the relative roles of innovators and everyone else, who he called 'imitators'. In a 1966 working paper, he predicted that new colour television sales – then still rising – would peak in 1968. 'Industry forecasts were much more optimistic than mine,' Bass later noted,[53] 'and it was perhaps to be expected that my forecast would not be well received.' Bass' prediction wasn't popular, but it ended up being much closer to reality. New sales indeed slowed then peaked, just as the model suggested they would.

As well as looking at how interest plateaus, we can also examine the early stages of adoption. When Everett Rogers published the S-curve in the early 1960s, he suggested that a new idea had 'taken off' once 20–25 per cent of people had adopted it. 'After that point, it is probably impossible to stop the further diffusion of a new idea,' he argued, 'even if one wishes to do so'. Based on outbreak dynamics, we can come up with a more precise definition for this take-off point. Specifically, we can work out when the number of new adoptions is growing fastest. After this point, a lack of susceptible people will start to slow the spread, causing the outbreak to eventually plateau. In Ross's simple model, the fastest growth occurs when just over 21 per cent of the potential audience have adopted the idea. Remarkably, this is the case regardless of how easily the innovation spreads.[54]

Ross's mechanistic approach is useful because it shows us what different types of happenings might look like in real life. Think about how the VCR adoption curve compares with the home ownership one: both eventually plateau, but the VCR curve grows exponentially at first. Simple models of contagion will usually predict this kind of growth, because each new adoption creates even more adoptions, whereas models of independent happenings will not. It doesn't mean that exponential growth is always a sign that something is contagious – there might be other reasons why people increasingly adopt a technology – but it does show how different infection processes can affect the shape of an outbreak.

If we think about the dynamics of an outbreak, we can also identify shapes that would be very unlikely in reality. Imagine a disease epidemic that increases exponentially until all of the population is affected. What would be required to generate this shape?

In large epidemics, transmission generally slows down because there aren't many susceptible people left to infect. For the epidemic to keep increasing faster and faster, infectious people would have to actively start seeking out the remaining susceptibles in the later stages of the epidemic. It's the equivalent of you catching a cold, finding all your friends who hadn't got it yet and deliberately coughing on them until they got infected. The most familiar scenario that would create this

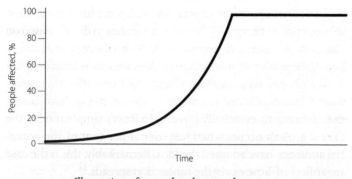

Illustration of an outbreak curve that grows
exponentially until everyone is affected

outbreak shape is therefore a fictional one: a group of zombies hunting down the last few surviving humans.

Back in real life, there are a few infections that affect their hosts in a way that increases transmission. Animals infected with rabies are often more aggressive, which helps the virus to spread through bites,[55] and people who have malaria can give off an odour that makes them more attractive to mosquitoes.[56] But such effects generally aren't large enough to overcome declining numbers of susceptibles in the later stages of an epidemic. What's more, many infections have the opposite effect on behaviour, causing lethargy or inactivity, which reduces the potential for transmission.[57] From innovations to infections, epidemics almost inevitably slow down as susceptibles become harder to find.

RONALD ROSS HAD PLANNED to study a whole range of outbreaks, but as his models became more complicated, the mathematics became trickier. He could outline the transmission processes, but he couldn't analyse the resulting dynamics. That's when he turned to Hilda Hudson, a lecturer at London's West Ham Technical Institute.[58] The daughter of a mathematician, Hudson had published her first piece of research in the journal *Nature* when she was ten years old.[59] She later studied at the University of Cambridge, where she was the only woman in her year to get first class marks in mathematics. Although she matched the results of the male student who ranked seventh, her performance wasn't included in the official listing (it wasn't until 1948 that women were allowed to receive Cambridge degrees[60]).

Hudson's expertise made it possible to expand the Theory of Happenings, visualising the patterns the different models could produce. Some happenings simmered away over time, gradually affecting everyone. Others rose sharply then fell. Some caused large outbreaks then settled down to a lower endemic level. There were outbreaks that came in steady waves, rising and falling with the seasons, and outbreaks that recurred sporadically. Ross and Hudson argued that the methods would cover most real-life situations. 'The rise and fall of epidemics as far as we can see at present can be explained by the general laws of happenings,' they suggested.[61]

Unfortunately, Hudson and Ross's work on the Theory of Happenings would be limited to three papers. One barrier was the First World War. In 1916, Hudson was called away to help design aircraft as part of the British war effort, work for which she would later get an OBE.[62] After the war, they faced another hurdle, with the papers ignored by their target audience. 'So little interest was taken in them by the "health authorities," that I have thought it useless to continue,' Ross later wrote.

When Ross first started working on the Theory of Happenings, he'd hoped it could eventually tackle 'questions connected with statistics, demography, public health, the theory of evolution, and even commerce, politics and statesmanship'.[63] It was a grand vision, and one that would eventually transform how we think about contagion. Yet even in the field of infectious disease research, several decades would pass before the methods became popular. And it would take even longer for the ideas to make their way into other areas of life.

Panics and pandemics

'I CAN CALCULATE THE MOTION of heavenly bodies but not the madness of people.' According to legend, Isaac Newton said this after losing a fortune investing in the South Sea Company. He'd bought shares in late 1719 and initially seen his investment rise, which persuaded him to cash in. However, the share price continued to climb and Newton – regretting his hasty sale – reinvested. When the bubble burst a few months later, he lost £20,000, equivalent to around £20 million in today's money.[1]

Great academic minds have a mixed record when it comes to financial markets. Some, like mathematicians Edward Thorp and James Simons, have set up successful investment funds, bringing in huge profits. Others have succeeded in sending money the opposite way. Take the hedge fund Long Term Capital Management (LTCM), which suffered massive losses following the Asian and Russian Financial Crises in 1997 and 1998. With two Nobel Prize-winning economists on its board and healthy initial profits, the firm had been the envy of Wall Street. Investment banks had lent them increasingly large sums of money to pursue increasingly ambitious trading strategies, to the point that when the fund went under in 1998, they had liabilities of over $100 billion.[2]

During the mid-1990s, a new phrase had become popular

among bankers. 'Financial contagion' described the spread of economic problems from one country to another. The Asian Financial Crisis was a prime example.[3] It wasn't the crisis itself that hit funds like LTCM; it was the indirect shockwaves that propagated through other markets. And because they'd lent so much to LTCM, banks also found themselves at risk. When some of Wall Street's most powerful bankers gathered on the tenth floor of the Federal Reserve Bank of New York on 23 September 1998, it was this fear of contagion that brought them there. To avoid LTCM's woes spreading to other institutions, they agreed a $3.6bn bailout. It was an expensive lesson, but unfortunately not one that was learned. Almost exactly ten years later, the same banks would be having the same conversations about financial contagion. This time it would be much worse.

I SPENT THE SUMMER of 2008 thinking about how to buy and sell the statistical concept of correlation. I'd just finished my penultimate year of university, and was interning with an investment bank in London's Canary Wharf. The basic idea was simple enough. Correlation measures how much things move in line with each other: if a stock market is highly correlated, stocks will tend to rise or fall together; if it's uncorrelated, some stocks might go up while others go down. If you think stocks are going to behave similarly in future, you'd ideally want a trading strategy that profited from this correlation. My job was to help develop such a strategy.

Correlation isn't just some niche topic to keep a mathematically minded intern occupied. It turns out to be crucial for understanding why 2008 would end with a full-blown financial crisis. It can also help explain how contagion spreads more generally, from social behaviour to sexually transmitted infections. As we'll see, it's a link that would eventually pull outbreak analysis into the heart of modern finance.

Each morning that summer, I took the Docklands Light Railway to work. Just before it reached my stop at Canary Wharf, the train would pass the skyscraper at 25 Bank Street. The building was home to Lehman Brothers. When I'd applied for internships in late 2007, Lehman had been one of the coveted destinations for many applicants. It was part of the elite 'bulge bracket' group of banks, which also included firms like Goldman Sachs, JP Morgan, and Merrill Lynch. Bear Stearns had been part of the club too, until its collapse in March 2008.

Bear, as the bankers called it, had gone under because of failed investments in the mortgage market. Soon after, JP Morgan bought the carcass for less than a tenth of its earlier value. By the summer, everyone in the industry was speculating on which firm would go under next. Lehman seemed to be top of the list.

For mathematics students, an internship in finance was the brightly lit path that distracted from all others. Everyone I knew on my degree course, regardless of their eventual career, signed up for one. I was about a month or so into my internship when I changed my mind, and decided to pursue a PhD instead of a job offer. A major factor was the course in epidemiology I'd taken earlier that year. I'd become fascinated by the idea that disease outbreaks didn't have to be this mysterious, unpredictable occurrence. With the right methods, we could pick them apart, uncover what was really going on, and hopefully do something about it.

But first, there was the question of what was going on around me in Canary Wharf. Despite having settled on another career path, I still wanted to understand what was happening to the banking industry. Why had rows of trading desks recently been emptied of their employees? Why were celebrated financial ideas suddenly crumbling? And how bad could it get?

I was based in equities, analysing company share prices, but in the preceding years the real money had been in credit-based investments. One investment stood out in particular: banks had increasingly bunched together mortgages and other loans into 'collateralized debt obligations' (CDOs). These products let investors take on some of the mortgage lender's risk and earn money in return.[4] Such approaches could be extremely lucrative. Sajid Javid, who in 2019 was appointed the UK's Chancellor of the Exchequer, reportedly earned around £3m a year trading various credit products before he left banking in 2009.[5]

CDOs were based on an idea borrowed from the life insurance industry. Insurers had noticed that people were more likely to die following the death of a spouse, a social effect known as 'broken heart syndrome'. In the mid-1990s, they developed a way to account for this effect when calculating insurance costs. It didn't take long for bankers to borrow the idea and find a new use for it. Rather than looking at deaths, banks were interested in what happened when someone defaulted on a mortgage. Would other households follow? Such borrowing of mathematical models is common in finance, as well as in other fields. 'Human beings have limited foresight and great imagination,' financial mathematician Emanuel Derman once noted, 'so that, inevitably, a model will be used in ways its creator never intended.'[6]

Unfortunately, the mortgage models had some major flaws. Perhaps the biggest problem was that they were based on historical house prices, which had risen for the best part of two decades. This period of history suggested that the mortgage market wasn't particularly correlated: if someone in Florida missed a payment, for example, it didn't mean someone in California would too. Although some had speculated that housing was a bubble set to burst, many remained optimistic. In July 2005, CNBC interviewed Ben Bernanke, who chaired President Bush's Council of Economic Advisers and would shortly become Chairman of the US Federal Reserve. What did Bernanke think the worst-case scenario was? What would happen if house prices dropped across the country? 'It's a pretty unlikely possibility,' Bernanke said.[7] 'We've never had a decline in house prices on a nationwide basis.'

In February 2007, a year before Bear Stearns collapsed, credit specialist Janet Tavakoli wrote about the rise of investment products like CDOs. She was particularly unimpressed with the models used to estimate correlations between mortgages. By making assumptions that were so far removed from reality, these models had in effect created a mathematical illusion, a way of making high-risk loans look like low-risk investments.[8] 'Correlation trading has spread through the psyche of the financial markets like a highly infectious thought virus,' Tavakoli noted. 'So far, there have been few fatalities, but several victims have fallen ill, and the disease is rapidly spreading.'[9] Others shared her skepticism, viewing popular correlation methods as an overly simplistic way of analysing mortgage products. One leading hedge fund reportedly kept an abacus in one of its conference rooms; there was a label next to it that read 'correlation model'.[10]

Despite the problems with these models, mortgage products remained popular. Then reality caught up, as house prices started to fall. During that 2008 summer, I came to the opinion that many had been aware of the potential implications. The investments were tumbling in value by the day, but it didn't seem to matter as long as there were still naïve investors out there to sell them on to. It was like carrying a sack of money that you know has a massive hole in the bottom, but not caring because you're stuffing so much more in the top.

As a strategy it was, well, full of holes. By August 2008, speculation was rife about just how empty the money bags were. Across the city, banks were looking for injections of funding, competing to court sovereign wealth funds in the Middle East. I remember equity traders grabbing passing interns to point out the latest drop in Lehman's share price. I'd walk past empty desks, where once profitable CDO teams had been let go. Some of my colleagues would glance up nervously whenever security walked by, wondering if they'd be next. The fear was spreading. Then came the crash.

THE RISE OF COMPLEX FINANCIAL PRODUCTS – and fall of funds like Long Term Capital Management – had persuaded central banks that they needed to understand the tangled web of financial trading. In May 2006 the Federal Reserve Bank of New York organised a conference to discuss 'systemic risk'. They wanted to identify factors that might affect the stability of the financial network.[11]

The conference attendees came from a range of scientific fields. One was ecologist George Sugihara. His lab in San Diego focused on marine conservation, using models to understand the dynamics of fish populations. Sugihara was also familiar with the world of finance, having spent four years working for Deutsche Bank in the late 1990s. During that period, banks had rapidly expanded their quantitative teams, seeking out people with experience of mathematical models. In an attempt to recruit Sugihara, Deutsche Bank had taken him on a luxury trip to a British country estate. The story goes that during dinner, a senior banker wrote a huge salary offer on a napkin. An astonished Sugihara didn't know what to say. Mistaking Sugihara's silence for disdain, the banker withdrew the napkin and proceeded to write an even bigger number. There was another pause, followed by another number. This time, Sugihara took the offer.[12]

Those years with Deutsche Bank would be highly profitable for both parties. Although the data involved financial stocks rather than fish stocks, Sugihara's experience with predictive models successfully transferred across to his new field. 'Basically, I modelled the fear and greed of mobs that trade,' he later told *Nature*.[13]

Another person to join the Federal Reserve discussions was Robert May, who had previously supervised Sugihara's PhD. An ecologist by training, May had worked extensively on analysis of infectious diseases. Although May was drawn into financial research largely by accident, he would go on to publish several studies looking at contagion in financial markets. In a 2013 piece for *The Lancet* medical journal, he noted the apparent similarity between disease outbreaks and financial bubbles. 'The recent rise in financial assets and the subsequent crash have rather precisely the same shape as the typical rise and fall of cases in an outbreak of measles or other infection,' he wrote. May pointed out that when an infectious disease epidemic rises it's bad news, and when it falls, it's good news. In contrast, it's generally seen as positive when financial prices rise and bad when they fall. But he argued that this is a false distinction: rising prices are not always a good sign. 'When something is going up without a convincing explanation about why it's going up, that really is an illustration of the foolishness of the people,' as he put it.[14]

One of the best-known historical bubbles is 'tulip mania', which gripped the Netherlands in the 1630s. In popular culture, it's a classic story of financial madness. Rich and poor alike poured more and more money into the flowers, to the point where tulip bulbs were going for the price of houses. One sailor who mistook a bulb for a tasty onion ended up in jail. Legend has it that when the market crashed in 1637, the economy suffered and some people drowned themselves in canals.[15] Yet according to Anne Goldgar at Kings College London, there wasn't really that much of a bulb bubble. She couldn't find a record of anybody who was ruined by the crash. Only a handful of wealthy people splashed out for the most expensive tulips. The economy was unharmed. Nobody drowned.[16]

Other bubbles have had a much larger impact. The first time that people used the word 'bubble' to describe overinflated investments was during the South Sea Bubble.[17] Founded in

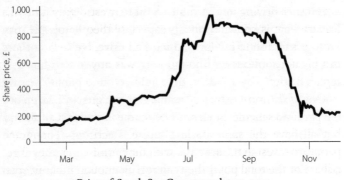

Price of South Sea Company shares, 1720
Data from Frehen et al., 2013[18]

1711, the British South Sea Company controlled several trading and slavery contracts in the Americas. In 1719, they secured a lucrative financial deal with the British government. The following year, the company's share price surged, rising four-fold in a matter of weeks, before falling just as sharply a couple of months later.[19]

Isaac Newton had sold most of his shares during the spring of 1720, only to invest again during the summer peak. According to mathematician Andrew Odlyzko, 'Newton did not just taste of the Bubble's madness, but drank deeply of it.' Some people timed their investments better. Bookseller Thomas Guy, an early investor, got out before the peak and used the profits to establish Guy's Hospital in London.[20]

There have been many other bubbles since, from Britain's Railway Mania in the 1840s to the US dot-com bubble in the late 1990s. Bubbles generally involve a situation where investors pile in, leading to a rapid rise in price, followed by a crash when the bubble bursts. Odlyzko calls them 'beautiful illusions', luring investors away from reality. During a bubble, prices can climb far above values that can be logically justified. Sometimes people invest simply on the assumption that more will join afterwards, driving up the value of their investment.[21] This can lead to what is known as the 'greater fool theory': people may know it's foolish to buy something expensive, but believe there is a greater fool out there, who will later buy it off them at a higher price.[22]

One of the most extreme examples of the greater fool theory is a pyramid scheme. Such schemes come in a variety of forms, but all have the same basic premise. Recruiters encourage people to invest in the scheme, with the promise that they'll get a share of the total pot if they can recruit enough other people. Because pyramid schemes follow a rigid format, they are relatively easy to analyse. Suppose a scheme starts with ten people paying in, and each of these people has to recruit ten others to get their payout. If they all manage to pull in another ten, it will mean 100 new people. Each of the new recruits will need to persuade another ten, which would grow the scheme by another 1,000 people. Expanding another step would require 10,000 extra people, then 100,000, then a million. It doesn't take long to spot that in the later stages of the scheme, there simply aren't enough people out there to persuade: the bubble will probably burst after a few rounds of recruitment. If we know how many people are susceptible to the idea, and might plausibly sign up, we can therefore predict how quickly the scheme will fail.

Given their unsustainable nature, pyramid schemes are generally illegal. But the potential for rapid growth, and the money it brings for the people at the top, means that they remain a popular option for scammers, particularly if there is a large pool of potential participants. In China, some pyramid schemes – or 'business cults' as the authorities call them – have reached a huge scale. Since 2010, several schemes have managed to recruit over a million investors each.[23]

Unlike pyramid schemes, which follow a rigid structure, financial bubbles can be harder to analyse. However, economist Jean-Paul Rodrigue suggests we can still divide a bubble into

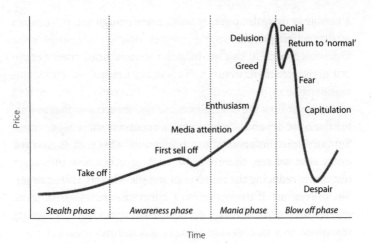

Price

Denial

Delusion

Return to 'normal'

Greed

Fear

Enthusiasm

Capitulation

Media attention

First sell off

Take off

Despair

Stealth phase   *Awareness phase*   *Mania phase*   *Blow off phase*

Time

The four phases of a bubble
Adapted from original graphic by Jean-Paul Rodrigue

four main stages. First, there is a stealth phase, where specialist
investors put money into a new idea. Next comes the awareness
phase, with a wider range of investors getting involved. There
may be an initial sell-off during this period as early investors cash
in, like Newton did in the early stages of the South Sea Bubble.
As the idea becomes more popular, the media and public join
in, sending prices higher and higher in a mania. Eventually the
bubble peaks and starts its decline during a 'blow off' phase,
perhaps with some small secondary peaks as optimistic investors hope for another rise. These bubble stages are analogous
to the four stages of an outbreak: spark, growth, peak, decline.[24]

One signature feature of a bubble is that it grows rapidly,
with the rate of buying activity increasing over time. Bubbles
often feature what's known as 'super-exponential' growth;[25] not
only does the buying activity accelerate, the acceleration itself
accelerates. With every increase in price, even more investors
join in, driving the price higher. And like an infection, the faster

a bubble grows, the faster it will burn through the population of susceptible people.

Unfortunately, it can be difficult to know how many people out there are still susceptible. This is a common problem when analysing an outbreak: during the initial growth phase, it's hard to work out how far through we are. For infectious disease outbreaks, a lot depends on how many infections show up as cases. Suppose most infections go unreported. This means that for every case we see, there will be a lot of other new infections out there, reducing the number of people who are still susceptible. In contrast, if the majority of infections are reported, there could still be a lot of people at risk of infection. One way around this problem is to collect and test blood samples from a population. If most people have already been infected and developed immunity to the disease, it's unlikely the outbreak can continue for much longer. Of course, it's not always possible to collect a large number of samples in a short space of time. Even so, we can still say something about the maximum possible outbreak size. By definition, it's impossible to have more infections than there are people in the population.

Things aren't so simple for financial bubbles. People can leverage their trades, borrowing money to cover additional investments. This makes it much harder to estimate how much susceptibility there is, and hence what phase of the bubble we're in. Still, it is sometimes possible to spot the signals of unsustainable growth. As the dot-com bubble grew in the late 1990s, a common justification for rising prices was the claim that internet traffic was doubling every 100 days. This explained why infrastructure companies were being valued at hundreds of billions of dollars and investors were pouring money into internet providers like WorldCom. But the claim was nonsense. In 1998, Andrew Odlyzko, then a researcher at AT&T labs, realised the internet was growing at a much slower rate, taking about a year to double in size.[26] In one press release, WorldCom had claimed that user demand was growing by 10 per cent every week. For this growth to be sustainable, it would mean that within a year or so, everyone in the world would have had to be active online for twenty-four hours a day.[27] There were simply not enough susceptible people out there.

Arguably the greatest bubble of recent years has been Bitcoin, which uses a shared public transaction record with strong encryption to create a decentralised digital currency. Or as comedian John Oliver described it: 'everything you don't understand about money combined with everything you don't understand about computers.'[28] The price of one Bitcoin climbed to almost $20,000 in December 2017, before dropping to less than a fifth of this value a year later.[29] It was the latest in a series of mini-bubbles; Bitcoin prices had risen and crashed several times since the currency emerged in 2009. (Prices would start to rise again in mid-2019.)

Each Bitcoin bubble involved a larger group of susceptible people, like an outbreak gradually making its way from a village into a town and finally into a city. At first, a small group of early investors got involved; they understood the Bitcoin technology and believed in its underlying value. Then a wider range of investors joined in, bringing more money and higher prices. Finally, Bitcoin hit the mass-market, with coverage on the front pages of newspapers and adverts on public transport. The delay between each of the historical Bitcoin peaks suggests that the idea didn't spread very efficiently between these different groups. If susceptible populations are strongly connected, an epidemic will generally peak around the same time, rather than as a series of smaller outbreaks.

According to Jean-Paul Rodrigue, there is a dramatic shift during the main growth phase of a bubble. The amount of money available increases, while the average knowledge base decreases. 'The market gradually becomes more exuberant as "paper fortunes" are made from regular "investors" and greed sets in,' he suggested.[30] Economist Charles Kindleberger, who wrote the landmark book *Manias, Panics, and Crashes* in 1978, along with Robert Aliber, emphasised the role of social contagion during this phase of a bubble: 'There is nothing so disturbing to one's well being and judgment as to see a friend get rich'.[31] Investors' desire to be part of a growing trend can even cause warnings about a bubble to backfire. During the British Railway Mania in the 1840s, newspapers like *The Times* argued that railway investment was growing too fast, potentially putting other parts of the economy at risk. But this only encouraged investors, who saw it as evidence that railway company stock prices would continue rising.[32]

In the later stages of a bubble, fear can spread in much the same way as enthusiasm. The first ripple in the 2008 mortgage bubble appeared as early as April 2006, when US house prices peaked.[33] It sparked the idea that mortgage investments were much riskier than people had thought, an idea that would spread through the industry, eventually bringing down entire banks in the process. Lehman Brothers would collapse on 15 September 2008, a week or so after I finished my internship in Canary Wharf. Unlike Long Term Capital Management, there would be no saviour. Lehman's collapse triggered fears that the entire global financial system could go under. In the US and Europe, governments and central banks provided over $14 trillion worth of support to prop up the industry. The scale of the intervention reflected how much banks' investments had expanded in the preceding decades. Between the 1880s and 1960s, British banks' assets were generally around half the size of the country's economy. By 2008, they were more than five times larger.[34]

I didn't realise it at the time, but as I was leaving finance for a career in epidemiology, in another part of London the two fields were coming together. Over on Threadneedle Street, the Bank of England was battling to limit the fallout from Lehman's collapse.[35] More than ever, it was clear that many had overestimated the stability of the financial network. Popular assumptions of robustness and resilience no longer held up; contagion was a much bigger problem than people had thought.

This is where the disease researchers came in. Building on that 2006 conference at the Federal Reserve, Robert May had started to discuss the problem with other scientists. One of them was Nim Arinaminpathy, a colleague at the University of Oxford. Arinaminpathy recalled that, pre-2007, it was unusual to study the financial system as a whole. 'There was a lot of faith in the vast, complex financial system being self-correcting,' he said. 'The attitude was "we don't need to know how the system works, instead we can concentrate on individual institutions".'[36] Unfortunately, the events of 2008 would reveal the weakness in this approach. Surely there was a better way?

During the late 1990s, May had been Chief Scientist to the UK Government. As part of this role, he'd got to know Mervyn King, who would later become Governor of the Bank of England. When the 2008 crisis hit, May suggested they look at the issue of contagion in more detail. If a bank suffered a shock, how might it propagate through the financial system? May and his colleagues were well placed to tackle the problem. In the preceding decades, they had studied a range of infections – from measles to HIV – and developed new methods to guide disease control programmes. These ideas would eventually revolution-ise central banks' approach to financial contagion. However, to understand how these methods work, we first need to look at a more fundamental question: how do we work out whether an infection – or a crisis – will spread or not?

AFTER WILLIAM KERMACK and Anderson McKendrick announced their work on epidemic theory in the 1920s, the field took a sharp mathematical turn. Although people continued working on out-break analysis, the work became more abstract and technical. Researchers like Alfred Lotka published lengthy, complicated papers, moving the field away from real-life epidemics. They found ways to study hypothetical outbreaks involving random events, intricate transmission processes and multiple populations. The emergence of computers helped drive these technical developments; models that were previously difficult to analyse by hand could now be simulated.[37]

Then progress stuttered. The obstacle was a 1957 textbook written by mathematician Norman Bailey. Continuing the theme of the preceding years, it was almost entirely theoretical, with hardly any real-life data. The textbook was an impressive survey of epidemic theory, which would help lure several young researchers into the field. But there was a problem: Bailey had left out a crucial idea, which would turn out to be one of the most important concepts in outbreak analysis.[38]

The idea in question had originated with George MacDonald, a malaria researcher based in the Ross Institute at the London School of Hygiene & Tropical Medicine. In the early 1950s, MacDonald had refined Ronald Ross's mosquito model, making it possible to incorporate real-life data about things like mosquito lifespan and feeding rates. By tailoring the model to actual scenarios, MacDonald was able to spot which part of the transmission process was most vulnerable to control measures. Whereas Ross had focused on the mosquito larvae that lived in water, MacDonald realised that to tackle malaria, agencies would be better off targeting the adult mosquitoes. They were the weakest link in the chain of transmission.[39]

In 1955, the World Health Organization announced plans to eradicate a disease for the first time. Inspired by MacDonald's analysis, they had chosen malaria. Eradication meant getting rid of all infections globally, something that would eventually prove harder to achieve than hoped; some mosquitoes became resistant to pesticides, and control measures targeting mosquitoes were less effective in some areas than others. As a result, WHO would later shift its focus to smallpox, eradicating the disease in 1980.[40]

MacDonald's idea to target adult mosquitos had been a crucial piece of research, but it wasn't the one that Bailey had omitted in his textbook. The truly groundbreaking idea had been nestled in the appendix of MacDonald's paper.[41] Almost as an afterthought, he had proposed a new way of thinking about infections. Rather than looking at critical mosquito densities, he suggested thinking about what would happen if a single infectious person arrived in the population. How many more infections would follow?

Twenty years later, mathematician Klaus Dietz would finally pick up on the idea in MacDonald's appendix. In doing so, he would help bring the theory of epidemics out of its mathematical niche and into the wider world of public health. Dietz outlined a quantity that would become known as the 'reproduction number', or R for short. R represented the number of new infections we'd expect a typical infectious person to generate on average.

In contrast to the rates and thresholds used by Kermack and McKendrick, R is a more intuitive – and general – way to think about contagion. It simply asks: how many people would we expect a case to pass the infection on to? As we shall see in later chapters, it's an idea that we can apply to a wide range of outbreaks, from gun violence to online memes.

R is particularly useful because it tells us whether to expect a large outbreak or not. If R is below one, each infectious person will on average generate less than one additional infection. We'd therefore expect the number of cases to decline over time. However, if R is above one, the level of infection will rise on average, creating the potential for a large epidemic.

Some diseases have a relatively low R. For pandemic flu, R is generally around 1–2, which is about the same as Ebola during the early stages of the 2013–16 West Africa epidemic. On average, each Ebola case passed the virus onto a couple of other people. Other infections can spread more easily. The SARS virus, which caused outbreaks in Asia in early 2003, initially had an R of 2–3. Its cousin

COVID-19, which spread widely in early 2020, also had an R of 2-3 when no control measures were in place.[42] Smallpox, which is still the only human infection that's been eradicated, had an R of 4–6 in an entirely susceptible population. Chickenpox is slightly higher, with an R around 6–8 if everyone is susceptible. Yet these numbers are low in comparison to what measles is capable of. In a fully susceptible community, a single measles case can generate more than 20 new infections on average.[43] Much of this is down to the incredible lingering power of the measles virus: if you sneeze in a room when you have the infection, there could still be virus floating around in the air a couple of hours later.[44]

As well as measuring transmission from a single infectious person, R can give clues about how quickly the epidemic will grow. Recall how the number of people in a pyramid scheme increased with each step. Using R, we can apply the same logic to disease outbreaks. If R is 2, an initial infected person will generate two cases on average. These two new cases will on average generate two more each, and so on. Carrying on doubling and by the fifth generation of the outbreak, we'd expect 32 new cases to appear; by the tenth, there would be 1,024 on average.

Because outbreaks often grow exponentially at first, a small change in R can have a big effect on the expected number of cases after a few generations. We've just seen that with an R of 2, we'd expect 32 new cases in the fifth generation of the outbreak. If R were 3 instead, we'd expect 243 at this same point.

One of the reasons R has become so popular is that it can be estimated from real-life data. From HIV to Ebola, R makes it possible to quantify and compare transmission for different diseases. Much of this popularity is down to Robert May and his long-standing collaborator Roy Anderson. During the late 1970s, the pair had helped bring epidemic research to a new audience. Both had a background in ecology, which gave them a more practical outlook than the mathematicians who'd preceded them. They were interested in data and how models could

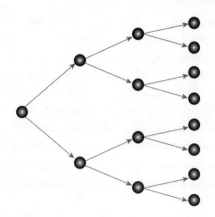

Example of an outbreak where each case infects two other
people. Circles are cases, arrows show route of transmission

apply to real-life situations. In 1980, May read a draft paper by
Paul Fine and Jacqueline Clarkson of the Ross Institute, who
had used a reproduction number approach to analyse measles
epidemics.[45] Realising its potential, May and Anderson quickly
applied the idea to other problems, encouraging others to join
them.

It soon became clear the reproduction number could vary a lot
between different populations. For example, diseases like measles
can spread to a lot of people if it hits a community with limited
immunity, but we rarely see outbreaks in countries with high
levels of vaccination. The R of measles can be 20 in populations
where everyone is at risk, but in highly vaccinated populations,
each infected person generates less than one secondary case on
average. In other words, R is below one in these places.

We can therefore use the reproduction number to work out
how many people we need to vaccinate to control an infection.
Suppose an infection has an R of 5 in a fully susceptible population, as smallpox did, but we then vaccinate four out of every
five people. Before vaccination, we'd have expected a typical

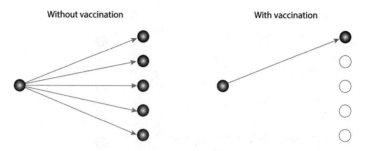

Comparison of transmission with and without 80 per cent
vaccination, when R is 5 in a fully susceptible population

infectious person to infect five other people. If the vaccine is 100
per cent effective, four of these people will now be immune on
average. So each infectious person would be expected to generate only one additional case.

If we instead vaccinate more than four fifths of the population, the average number of secondary cases will drop below
one. We'd therefore expect the number of infections to decline
over time, which would bring the disease under control. We
can use the same logic to work out vaccination targets for other
infections. If R is 10 in a fully susceptible population, we'd need
to vaccinate at least 9 in every 10 people. If R is 20, as it can be
for measles, we need to vaccinate 19 out of every 20, or over
95 per cent of the population, to stop outbreaks. This percentage is commonly known as the 'herd immunity threshold'. The
idea follows from Kermack and McKendrick's work: once this
many people are immune, the infection won't be able to spread
effectively.

Reducing the susceptibility of a population is perhaps the
most obvious way to bring down the reproduction number, but
it's not the only one. It turns out that there are four factors that
influence the value of R. Uncovering them is the key to understanding how contagion works.

ON 19 APRIL 1987, Princess Diana opened a new treatment unit in London's Middlesex Hospital. While there, she did something that surprised the accompanying media and even the hospital staff: she shook a patient's hand. The unit was first in the country that was specifically built to care for people with AIDS. The handshake was significant because despite scientific evidence the disease could not spread through touch, there was still a common belief that it could.[46]

The rise of HIV/AIDS in the 1980s created an urgent need to uncover how the epidemic was spreading. What features of the disease were driving transmission? The month before Diana visited Middlesex Hospital, Robert May and Roy Anderson had published a paper that broke down the reproduction number for HIV.[47] They noted that R was influenced by a number of different things. First, it depends on how long a person is infectious: the shorter an infection is, the less time there is to give it to someone else. As well as the duration of infection, R will depend on how many people someone interacts with while infectious. If they have a lot of contact with others, it will provide plenty of opportunities for the infection to spread. Finally, it depends on the probability that the infection is passed on during each of these encounters, assuming the other person is susceptible.

R therefore depends on four factors: the *duration* of time a person is infectious; the average number of *opportunities* they have to spread the infection each day they're infectious; the probability an opportunity results in *transmission*; and the average *susceptibility* of the population. I like to call these the 'DOTS' for short. Joining them together gives us the value of the reproduction number:

R = **D**uration × **O**pportunities × **T**ransmission probability × **S**usceptibility

Breaking the reproduction number down into these DOTS components, we can see how different aspects of transmission trade off against each other. This can help us work out the best way to control an epidemic, because some aspects of the reproduction number will be easier to change than others. For example, widespread sexual abstinence would reduce the number of *opportunities* for HIV transmission, but it's not an appealing or practical option for most people. Health agencies have therefore focused on getting people to use condoms, which reduce the probability of *transmission* during sex. In recent years, there has also been a lot of success with so-called pre-exposure prophylaxis (PrEP), whereby HIV-negative people take anti-HIV drugs to reduce their *susceptibility* to the infection.[48]

The type of transmission opportunities we're interested in will depend on the infection. For HIV and gonorrhea, transmission occurs mostly through sexual encounters, while infections like COVID-19 and smallpox can spread during face-to-face conversations. But the same overall idea applies. In the example of COVID-19, if people self-isolate when they have symptoms, it in effect reduces *duration*; limiting gathering size reduces *opportunities*; masks or physical distancing can reduce the *transmission* probability; and immunity following infection – or vaccination, if available – could reduce *susceptibility*.

The trade-off in the DOTS means that if someone is infectious for twice as long, in transmission terms it's equivalent to them making twice as many contacts. In the past, smallpox and HIV have at times both had an R of around 5.[49] However, people are generally infectious with smallpox for a shorter period, which means that there must be more opportunities to spread infection per day, or a higher transmission probability during each opportunity, to compensate.

The reproduction number has become a crucial part of modern outbreak research, but there's another feature of contagion we also need to consider. Because R looks at the average level of transmission, it doesn't capture some of the unusual events that can occur during outbreaks. One such event happened in March

1972, when a Serbian teacher arrived at Belgrade's main hospital with an unusual mix of symptoms. He'd been given penicillin at his local medical centre to treat a rash, but severe haemorrhaging had followed. Dozens of students and staff in the hospital gathered to see what they presumed was a strange reaction to the drug. But it was no allergy. After the man's brother also fell ill, staff realised what the real problem was, and what they had exposed themselves to. The man had been infected with smallpox, and there would be 38 more cases – all traceable to him – before the infections in Belgrade subsided.[50]

Although smallpox wouldn't be eradicated globally until 1980, it was already gone from Europe, with no cases reported in Serbia since 1930. The teacher had likely caught the disease from a local clergyman who'd recently returned from Iraq. Several similar flare-ups had happened in Europe during the 1960s and 1970s, most of them travel-related. In 1961, a girl returned from Karachi, Pakistan to Bradford, England, bringing the smallpox virus with her and unwittingly infecting ten other people. An outbreak in Meschede, Germany, in 1969 also started with a visitor to Karachi. This time it was a German electrician who'd travelled there; he would pass the infection on to seventeen others.[51] However, these events weren't typical: most cases who returned to Europe didn't infect anyone.

In a susceptible population, smallpox has a reproduction number of around 4–6. This represents the number of secondary cases we'd expect to see, but it's still just an average value: in reality there can be a lot of variation between individuals and outbreaks. Although the reproduction number provides a useful summary of overall transmission, it doesn't tell us how much of this transmission comes from a handful of what epidemiologists call 'superspreading' events.

A common misconception about disease outbreaks is that they grow steadily generation-by-generation, with each case infecting a similar number of people. If an infection spreads from person-to-person, creating a chain of cases, we refer to it as 'propagated transmission'. However, propagated outbreaks don't necessarily follow the clockwork pattern of the reproduction number, growing by the exact same amount each generation. In 1997, a group of epidemiologists proposed the '20/80 rule' to describe disease transmission. For diseases like HIV and malaria they'd found that 20 per cent of cases were responsible for around 80 per cent of transmission.[52] But like most biological rules, there were some exceptions to the 20/80 rule of transmission. The researchers had focused on sexually transmitted infections (STIs) and mosquito-borne infections. Other outbreaks didn't always follow this pattern. After the 2003 SARS epidemic – which had involved several instances of mass infection – there was renewed interest in the notion of superspreading. For SARS, it seemed to be particularly important: 20 per cent of cases caused almost 90 per cent of transmission. In early 2020, our group and others estimated that COVID-19 was similar.[53] In contrast, diseases like plague have fewer superspreading events, with the top 20 per cent of cases responsible for only 50 per cent of transmission.[54]

In other situations, an outbreak may not be propagated at all. It may be the result of 'common source transmission', with all cases coming from the same place. One example is food poisoning: outbreaks can often be traced to a specific meal or person. The most infamous case is that of Mary Mallon – often referred to as 'Typhoid Mary' – who carried a typhoid infection without symptoms. In the early twentieth century, Mallon was employed as a cook for several families around New York City, leading to multiple outbreaks of the disease and several deaths.[55]

During a common source outbreak, cases often appear within a short period of time. In May 1916, there was a typhoid outbreak in California a few days after a school picnic. Like Mallon, the cook who'd made the ice cream for the picnic had been carrying the infection without knowing.

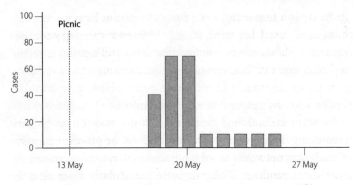

Typhoid outbreak following a picnic in California, 1916[56]

We can therefore think of disease transmission as a contin-
uum. At one end, we have a situation where a single person
– such as Mary Mallon – generates all of the cases. This is the
most extreme example of superspreading, with one source
responsible for 100 per cent of transmission. At the other end,
we have a clockwork epidemic where each case generates
exactly the same number of secondary cases. In most cases, an
outbreak will lie somewhere between these two extremes.

If there is potential for superspreading events during an
outbreak, it implies that some groups of people might be particularly important. When researchers realised that 80 per cent
of HIV transmission came from 20 per cent of cases, they suggested targeting control measures at these 'core groups'. For
such approaches to be effective, though, we need to think about
how individuals are connected in a network – and why some
people might be more at risk than others.

THE MOST PROLIFIC MATHEMATICIAN in history was an academic
nomad. Paul Erdős spent his career travelling the world, living
from two half-full suitcases without a credit card or chequebook.

'Property is a nuisance,' as he put it. Far from being a recluse, though, he used his trips to accumulate a vast network of research collaborations. Fuelled by coffee and amphetemines, he'd turn up at colleagues' houses, announcing that 'my brain is open'. By the time he died in 1996, he'd published about 1,500 papers, with over eight thousand co-authors.[57]

As well as building networks, Erdős was interested in researching them. Along with Alfréd Rényi, he pioneered a way of analysing networks in which individual 'nodes' were linked together at random. The pair were particularly interested in the chance these networks would end up being fully connected – with a possible route between any two nodes – rather than split into distinct pieces. Such connectedness matters for outbreaks. Suppose a network represents sexual partnerships. If it's fully connected, a single infected person could in theory spread an STI to everyone else. But if the network is split into many pieces, there's no way for a person in one component to infect somebody in another.

It can also make a difference if there is a single path across the network, or several. If networks contain closed loops of contacts, it can increase STI transmission.[58] When there's a loop, the infection can spread across the network in two different ways;

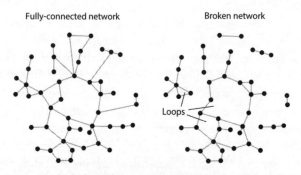

Illustration of fully-connected and broken Erdős–Rényi networks

even if one of the social links breaks, there's still another route left. For STIs, outbreaks are therefore more likely to spread if there are several loops present in the network.

Although the randomness of Erdős–Rényi networks is convenient from a mathematical point of view, real life can look very different. Friends cluster together. Researchers collaborate with the same group of co-authors. People often have only one sexual partner at a time. There are also links that go beyond such clusters. In 1994, epidemiologists Mirjam Kretzschmar and Martina Morris modelled how STIs might spread if some people had multiple sexual partners at the same time. Perhaps unsurprisingly, they found that these partnerships could lead to a much faster outbreak, because they created links between very different parts of the network.

The Erdős–Rényi model could capture the occasional long-range connections that occurred in real networks, but it couldn't reproduce the clustering of interactions. This discrepancy was resolved in 1998, when mathematicians Duncan Watts and Steven Strogatz developed the concept of a 'small-world' network, in which most links were local but a few were long-range. They found that such networks cropped up in all sorts of places: the electricity grid, neurons in worm brains, co-stars in film casts, even Erdős's academic collaborations.[59] It was a remarkable finding, and more discoveries were about to follow.

The small-world idea had addressed the issue of clustering and long-range links, but physicists Albert-László Barabási and Réka Albert spotted something else unusual about real-life networks. From film collaborations to the World Wide Web, they'd noticed that some nodes in the network had a huge number of connections, far more than typically appeared in the Erdős–Rényi or small-world networks. In 1999, the pair proposed a simple mechanism to explain this extreme variability in connections: new nodes that joined the network would preferentially attach to already popular ones.[60] It was a case of the 'rich get richer'.

The following year, a team at the University of Stockholm showed that the number of sexual partnerships in Sweden also appeared to follow this rule: the vast majority of people had slept with at most one person in the past year, whereas some reported dozens of partners. Researchers have since found similar patterns of sexual behaviour in countries ranging from Burkina Faso to the United Kingdom.[61]

What effect does this extreme variability in number of partners have on outbreaks? In the 1970s, mathematician James Yorke and his colleagues noticed there was a problem with the ongoing gonorrhea epidemic in the United States. Namely, it didn't seem possible. For the disease to keep spreading, the reproduction number needed to be above one. That meant infected people should on average have at least two recent sexual partners: one who gave the infection to them, and another who they passed it on to. But a study of patients with gonorrhea had found that they'd had only 1.5 recent partners on average.[62] Even if the probability of transmission during sex was very high, it suggested that there simply weren't enough encounters for the disease to persist. What was going on?

If we just take the average number of partners, we are ignoring the fact that not everyone's sex lives are the same. This variability is important: if someone has a lot of partners, we'd expect them to be both more likely to get infected and more likely to pass the infection on. We therefore need to account for the fact that they can contribute to transmission in these two different ways. Yorke and his colleagues argued that this might explain why there could be a gonorrhea epidemic, despite people having few partnerships on average: people with lots of contacts might be contributing disproportionality to the spread, pushing the reproduction number above one. Anderson and May would later show that the more variation there was in the number of partners people had, the higher we'd expect the reproduction number to be.

Identifying people who are at higher risk – and finding ways to reduce this risk – can help stop an outbreak in its early stages. In the late 1980s, Anderson and May suggested that STIs would initially spread quickly through such high-risk groups, even though the overall outbreak would be smaller than we'd expect if everyone mixed at random.[63]

By breaking contagion down into its basic DOTS components – duration, opportunities, transmission probability, susceptibility – and thinking about how network structure affects contagion, we can also estimate the risk posed by a new STI. In 2008, an American scientist returned home to Colorado after a month working in Senegal. A week later, he'd fallen ill with a headache, extreme tiredness, and a rash on his torso. Soon after, his wife – who hadn't travelled – developed the same symptoms. Subsequent lab tests indicated both had been exposed to the Zika virus. Prior Zika research had focused on transmission from mosquitoes, but the Colorado incident suggested the virus had access to another route: it could infect people during sexual encounters.[64] As Zika spread around the globe in 2015–16, more reports of sexual transmission would follow, fueling speculation about a new type of outbreak. 'Zika: The Millennials' S.T.D.?' asked one opinion piece in the *New York Times* during 2016.[65]

Based on the DOTS for Zika, our research group estimated that the reproduction number for sexual transmission was below one; the virus would probably not cause an STI epidemic. Zika could potentially cause small outbreaks in groups with a lot of sexual contacts, but it was unlikely to pose a major risk in areas without mosquitoes.[66] Unfortunately, the same has not been true for other STIs.

GAËTAN DUGAS WAS BLOND, charming, and had a lot of sex. A Canadian flight attendant, he'd slept with over two hundred men a year prior to March 1984, when he died of AIDS a few weeks after his 31st birthday. Three years later, journalist Randy Shilts featured Dugas in his bestselling book *And the Band Played On*. Shilts suggested that Dugas had played a central part in the early spread of the disease. He dubbed Dugas 'patient zero', a term still used today to refer to the first case in an outbreak. Shilts' book fuelled speculation that Dugas was the person who introduced the epidemic to North America. The *New York Post* called him 'The Man Who Gave Us AIDS'; the *National Review* said he was 'the Columbus of AIDS'.

The idea of Dugas as patient zero was certainly attention-grabbing, and has been repeated often in the decades since. But it turned out to be fiction. In 2016, a team of researchers published an analysis of HIV viruses from a range of patients, including men diagnosed with AIDS in the 1970s and Dugas himself. Based on the genetic diversity of these viruses and the rate of HIV evolution, the team estimated that HIV had arrived into North America in 1970 or 1971. However, they found no evidence that Dugas had introduced HIV to the US. He was just another case in a much wider epidemic.[67]

So how did the patient zero designation come about? In the original outbreak investigation, Dugas hadn't actually been listed as 'Patient o', but rather as 'Patient O', the 'O' short for 'Outside California'. In 1984, William Darrow, a researcher with the Centers for Disease Control and Prevention (CDC), had been assigned to investigate a cluster of deaths among gay men in Los Angeles.[68] The CDC generally gave each case a number based on the order in which they had been reported, but the cases had been relabelled for the LA analysis. Before Dugas had been linked to the Los Angeles cluster, he was simply 'Patient 057'.

When investigators traced how the cases were linked, it suggested that the deaths might be the result of an as-yet-unknown STI. Dugas appeared prominently in the network, with links to multiple cases in New York and LA. This was in part because he'd tried to help the investigators, naming 72 of his partners in the preceding three years. Darrow pointed out that this had always been the aim of the investigation: to understand how cases were linked, rather than find out who had started the outbreak. 'I never said that he was the first case in the United States,' he later commented.

When investigating outbreaks, we face a gap between what we want to know and what we can measure. Ideally, we'd have data on all the ways in which people are connected, and how infection has spread through these links. What we can actually measure is very different. A typical outbreak investigation will reconstruct some of the links between people who were infected. Depending on which cases and links are reported, the resulting network won't necessarily look like the actual transmission route. Some people might appear more prominent than they really were, while some transmission events might be missed.

When Randy Shilts came across the CDC diagram while researching his book, his attention was drawn to Dugas. 'In the middle of that study was a circle with an O next to it, and I always thought it was Patient O,' he later recalled. 'When I went to the CDC, they started talking about Patient Zero. I thought, "Ooh, that's catchy".'[69]

It's easier to tell a story when it has a clear antagonist. According to historian Phil Tiemeyer, it was Shilts's editor Michael Denneny who suggested they make Dugas the villain in the book and accompanying publicity. 'Randy hated the idea,' Denneny told Tiemeyer. 'It took me almost a week to argue him into it.' The decision – which Denneny later said he regretted – came because the media seemed to have little interest in AIDS otherwise. 'They were not going to review a book that was an indictment of the Reagan administration and the medical establishment.'[70]

When discussing outbreaks that involve superspreading

events, there is a tendency to place all attention on the people apparently at the centre of them. Who are these 'superspreaders'? What makes them different from everyone else? However, such attention can be misplaced. Take that story of the Belgrade teacher who arrived in hospital with smallpox. There was nothing intrinsically unusual about him or his behaviour. He had acquired the disease through a chance encounter, had tried to get medical care at an appropriate place – a hospital – and the outbreak spread because nobody initially suspected smallpox was the cause. This is true of many outbreaks: it's often difficult to predict in advance what role a specific individual will play.

Even if we can identify situations that create a risk of disease transmission, it won't necessarily lead to the outcome we expect. On 21 October 2014, at the height of the Ebola epidemic in West Africa, a two-year-old girl arrived at a hospital in the city of Kayes, Mali. Following the death of her father, who had been a healthcare worker, the girl had travelled over 1,200 km from neighbouring Guinea with her grandmother, uncle and sister. At the Kayes hospital, the girl tested positive for Ebola, and would die of the disease the next day. She was Mali's first case of Ebola, and health authorities began to search for people who may have come into contact with her. During her trip, she'd taken at least one bus and three taxis, potentially interacting with dozens if not hundreds of people. She'd already been displaying symptoms when she arrived at the hospital; based on the nature of Ebola transmission, there was a good chance she could have passed the virus on. Investigators eventually managed to track down over one hundred of the girl's contacts and placed them in quarantine as a precaution. However, none of them came down with Ebola. Despite her long journey, the girl hadn't infected anyone.[71]

When Ebola superspreading events did occur during 2014–15, our team noticed there was one feature that stood out. Unfortunately, it wasn't a particularly helpful one: the cases most likely to be involved in superspreading were the ones that couldn't be linked to existing chains of transmission. Put simply, the people driving the epidemic were generally the ones the health authorities didn't know about. These people went undetected until they sparked a new set of infections, making it near impossible to predict superspreading events.[72]

With enough effort, we can often trace some of the path of infection during an outbreak, reconstructing who might have infected whom. It can be tempting to construct a narrative as well, speculating about why certain people transmitted more than others. However, just because an infection is capable of superspreading doesn't necessarily mean the same people are always the superspreaders. Two people might behave in almost the same way, but by chance one of them spreads infection and the other does not. When history is written, one is blamed and the other ignored. Philosophers call it 'moral luck': the idea that we tend to view actions with unfortunate consequences as worse than equal actions without any repercussions.[73]

Sometimes the people involved in an outbreak do behave differently, but not necessarily in the way we might assume. In his book *The Tipping Point*, Malcolm Gladwell describes an outbreak of gonorrhea in Colorado Springs, Colorado, during 1981. As part of the outbreak investigation, epidemiologist John Potterat and his colleagues had interviewed 769 cases, asking whom they'd recently had sexual contact with. Of these cases, 168 people had at least two contacts who were also infected. This suggested they were disproportionately important in the outbreak. 'Who were those 168 people?' Gladwell asked. 'They aren't like you or me. They are people who go out every night, people who have vastly more sexual partners than the norm, people whose lives and behaviour are well outside of the ordinary.'

Were these people really so promiscuous and unusual? Not particularly, in my view: the researchers found that, on average, these cases reported sexual encounters with 2.3 other infected people. This implies they caught the infection from one person and typically gave it to one or two others. Cases tended to be black or Hispanic, young, and associated with the military; almost half had known their sexual partners for more than two months.[74] During the 1970s, Potterat had begun to notice that promiscuity wasn't a good explanation for gonorrhea outbreaks in Colorado Springs. 'Especially striking was the difference in gonorrhea test outcome between sexually adventurous white women from a local upper middle class college and similarly aged black women with modest sexual histories and educational backgrounds,' he noted.[75] 'The former were seldom diagnosed with gonorrhea, unlike the latter.' A closer look at the Colorado Springs data suggested that transmission was likely to be the result of delays in getting treatment among certain social groups, rather than an unusually high level of sexual activity.

Viewing at-risk people as special or different can encourage a 'them and us' attitude, leading to segregation and stigma. In turn, this can make epidemics harder to control. From HIV/AIDS to Ebola, blame – and fear of blame – has pushed many outbreaks out of view. Suspicion around disease can result in many patients and their families being shunned by the local community.[76] This makes people reluctant to report the disease, which in turn amplifies transmission, by making the most important individuals harder to reach.

In early February 2020, a British man who had travelled to Singapore was dubbed a COVID-19 'superspreader' after subsequently passing infection on to several others at a ski chalet in France, before returning to the UK. A few weeks later, with the UK reporting its first local transmission, the media again latched onto the idea of finding someone to blame. 'Coronavirus: hunt for Patient Zero, Britain's virus spreader,' read the headline in the *Sunday Times* on 1 March.[77] How many people, upon reading such headline, would want to be that patient? How many, as a result, would avoid getting a test for that mild cough they had had since returning from a recent ski trip to northern Italy?

Unfortunately, the hunting mindset moved beyond head-lines. Reports of 'Wuhan coronavirus' and 'Chinese flu' were followed by reports of racially motivated attacks, from London to Los Angeles. Several prominent individuals, including US politicians and media hosts, have since suggested that terms like 'Chinese flu' are justified: if China had reported the extent of outbreak earlier, the argument goes, the US could have prepared adequately. And yet, by early March – five weeks after WHO declared the growing epidemic a Public Health Emergency of International Concern – an investigation by *The Atlantic* could only verify that 1,895 people in the US had been tested for the virus (for context, the UK had tested over 20,000 people by that point).[78] As happened in the early stages of the AIDS pandemic, some politicians and media outlets preferred to blame the crisis on a specific community, rather than face up to shortcomings closer to home.

Blaming certain groups for outbreaks is not a new phenom-enon. In the sixteenth century, the English believed syphilis came from France, so referred to it as the 'French pox'. The French, believing it to be from Naples, called it the 'Neopolitan disease'. In Russia, it was the Polish disease, in Poland it was Turkish, and in Turkey it was Christian.[79]

Such blame can stick for a long time. We still refer to the 1918 influenza pandemic, which killed tens of millions of people globally, as the 'Spanish flu'. The name emerged during the outbreak because media reports suggested Spain was the worst hit country in Europe. However, these reports weren't quite what they seemed. At the time, Spain had no wartime cen-sorship of news reports, unlike Germany, England and France, who quashed news of disease for fear that it might damage morale. The media blackout in these countries therefore made it appear that Spain had far more cases than anywhere else. (For their part, the Spanish media tried to blame the disease on the French.[80])

If we want to avoid country-specific disease names, it helps to suggest an alternative. One Saturday morning in March 2003, a group of experts gathered at WHO headquarters in Geneva to discuss a newly discovered infection in Asia.[81] Cases had already appeared in Hong Kong, China and Vietnam, with another reported in Frankfurt that morning. WHO was about to announce the threat to the world, but first they needed a name. They wanted something that was easy to remember, but which wouldn't stigmatize the countries involved. Eventually they settled on 'Severe Acute Respiratory Syndrome', or SARS for short.

THE SARS EPIDEMIC WOULD RESULT in over eight thousand cases and several hundred deaths, across multiple continents. Despite being brought under control in June 2003, the epidemic would cost an estimated $40 billion dollars globally.[82] It wasn't just the direct cost of treating disease cases; it was the economic impact of closed workplaces, empty hotels and cancelled trade.

According to Andy Haldane, now Chief Economist at the Bank of England, the wider effects of the SARS epidemic were comparable with the fallout from the 2008 financial crisis. 'These similarities are striking,' he said in a 2009 speech.[83] 'An external event strikes. Fear grips the system, which, in consequence, seizes. The resulting collateral damage is wide and deep.'

Haldane suggested that the public typically respond to an outbreak in one of two ways: flight or hide. In the case of an infectious disease, flight means trying to leave an affected area in the hope of avoiding infection. Because of travel restrictions and other control measures, this generally wasn't an option during the SARS epidemic.[84] Had infected people travelled – rather than being identified and isolated by health authorities – it could have spread the virus to even more locations. The flight response can also happen in finance. Faced with a crash, investors may cut their losses and sell off assets, driving prices even lower.

Alternatively, people may 'hide' during an outbreak, dodging situations that could potentially bring them into contact with the infection. If it's a disease outbreak, they might wash their hands more often, or reduce their social interactions. In finance, banks might hide by hoarding money rather than risking lending to other institutions. However, Haldane pointed out that there is a crucial difference between hide responses in disease outbreaks and financial crises. Hiding behaviour will generally help reduce disease transmission, even if it incurs a cost in the process. In contrast, when banks hoard money it can amplify problems, as happened with the 'credit crunch' that hit economies in the run up to the 2008 crisis.

Although the notion of a credit crunch would make headlines during 2007/8, economists first coined the term back in 1966. That summer, US banks had abruptly stopped lending. In the preceding years, there had been high demand for loans, with banks making more and more credit available to keep pace. Eventually, it had got to the point where banks weren't taking in enough money in savings to continue lending, so the loans stopped. It wasn't just a matter of banks asking borrowers for higher interest rates. They weren't lending at all. Banks had reduced the availability of loans before – there were several instances of 'credit squeezes' in the US during the 1950s – but some thought 'squeeze' was too gentle a word to describe the sudden impact of 1966. 'A "crunch" is different,' wrote economist Sidney Homer at the time. 'It is painful by definition, and it can even break bones.'[85]

The 2008 crisis wasn't the first time Andy Haldane had thought about contagion in financial systems.[86] 'I remember back in 2004/5, writing a note about us having entered the era of "super-systemic risk" as a result of these sorts of infections.'

His note suggested that the financial network might be robust in some situations and extremely fragile in others. The idea was well-established in ecology: the structure of a network might make it resilient to minor shocks, but the same structure could also leave it vulnerable to complete collapse if put under enough stress. Think about a team at work. If most people are doing well, weaker members can get away with mistakes because they are linked to high performers. However, if most of the team are struggling, the same links will instead drag strong members down. 'The basic point was that all this integration did indeed reduce the probability of mini-crashes,' Haldane said, 'but increased the probability of a maxi-crash.'

It may have been a prescient idea, but it didn't spread very far. 'That note didn't really go anywhere unfortunately,' he said, 'until the big one came.' Why didn't the idea take off? 'It was hard to spot any examples of such systemic risk at the time. It appeared to be a very flat ocean at that point.' That would change in autumn 2008. After Lehman Brothers collapsed, people across the banking industry started thinking in terms of epidemics. According to Haldane, it was the only way to explain what had happened. 'You couldn't tell a story about why Lehman had brought the financial system down without telling a contagion story.'

IF YOU WERE TO MAKE a list of network features that could amplify contagion, you'd find that the pre-2008 banking system had most of them. Let's start with the distribution of links between banks. Rather than connections being scattered evenly, a handful of firms dominated the network, creating massive potential for superspreading. In 2006, researchers working with the Federal Reserve Bank of New York picked apart the structure of the US Fedwire payment network. When they looked at the $1.3 trillion of transfers that happened between thousands of US banks on a

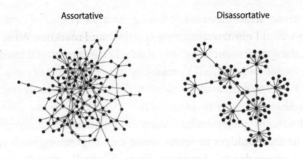

Illustration of assortative and disassortative networks
Adapted from Hao et al., 2011, PLOS ONE

typical day, they found that 75 per cent of the payments involved just 66 institutions.[87]

The variability in links wasn't the only problem. It was also how these big banks fitted into the rest of the network. In 1989, epidemiologist Sunetra Gupta led a study showing that the dynamics of infections could depend on whether a network is what mathematicians call 'assortative' or 'disassortative'. In an assortative network, highly connected individuals are linked mostly to other highly connected people. This results in an out-break that spreads quickly through these clusters of high-risk individuals, but struggles to reach the other, less connected, parts of the network. In contrast, a disassortative network is when high-risk people are mostly linked to low risk ones. This makes the infection spread slower at first, but leads to a larger overall epidemic.[88]

The banking network, of course, turned out to be disassorta-tive. A major bank like Lehman Brothers could therefore spread contagion widely; when Lehman failed, it had trading relation-ships with over one million counter-parties.[89] 'It was entangled in this mesh of exposures – derivatives and cash – and no one had the faintest idea quite who owed what to whom,' Haldane said. It didn't help that there were numerous, often hidden,

loops in the wider network, creating multiple routes of transmission from Lehman to other companies and markets. What's
more, these routes could be very short. The international financial network had become a smaller world during the 1990s and
2000s. By 2008, each country was only a step or two away from
another nation's crisis.[90]

In February 2009, investor Warren Buffett used his annual
letter to shareholders to warn about the 'frightening web of
mutual dependence' between large banks.[91] 'Participants
seeking to dodge troubles face the same problem as someone
seeking to avoid venereal disease,' he wrote. 'It's not just whom
you sleep with, but also whom they are sleeping with.' As well
as putting supposedly careful institutions at risk, Buffett suggested that the network structure could also incentivise bad
behaviour. If the government needed to step in and help during
a crisis, the first companies on the list would be those that were
capable of infecting many others. 'Sleeping around, to continue
our metaphor, can actually be useful for large derivatives dealers
because it assures them government aid if trouble hits.'

Given the apparent vulnerability of the financial network,
central banks and regulators needed to understand the 2008
crisis. What else had been driving transmission? The Bank of
England had already been working on models of financial contagion pre-crisis, but 2008 brought a new, real-life urgency to
the work. 'We started using them in practice when the crisis
broke,' Haldane said. 'Not just for making sense of what was
going on, but more importantly for what we might do to stop
it happening again.'

WHEN ONE BANK LENDS money to another, it creates a tangible
link between the two: if the borrower goes under, the lender
loses their money. In theory, we could trace this network to
understand the outbreak risk, just as we can for STIs. But there's

more to it than that. Nim Arinaminpathy has pointed out that networks of loans were just one of several problems in 2008. 'It's almost like HIV,' he said. 'You can have transmission through sexual contacts, as well as needle exchanges or blood transfusions. There are multiple routes of transmission.' In finance, contagion can also come from several different sources. 'It isn't just lending relationships, it's also about shared assets and other exposures.'

A long-standing idea in finance is that banks can use diversification to reduce their overall risk. By holding a range of investments, individual risks will balance each other out, improving the bank's stability. In the lead up to 2008, most banks had adopted this approach to investment. They'd also chosen to do it in the same way, chasing the same types of assets and investment ideas. Although each individual bank had diversified their investments, there was little diversity in the way they had collectively done it.

Why the similarity in behaviour? During the Great Depression that followed the 1929 Wall Street crash, economist John Maynard Keynes observed that there is a strong incentive to follow the crowd. 'A sound banker, alas, is not one who foresees danger and avoids it,' he once wrote, 'but one who, when he is ruined, is ruined in a conventional way along with his fellows, so that no one can really blame him.'[92] The incentive works the other way too. Pre-2008, many companies started investing in trendy financial products like CDOs, which were far outside their area of expertise. Janet Tavakoli has pointed out that banks were happy to indulge them, inflating the bubble further. 'As they say in poker, if you don't know how to spot the sucker at the table, it is you.'[93]

When multiple banks invest in the same asset, it creates a potential route of transmission between them. If a crisis hits and one bank starts selling off its assets, it will affect all the other firms who hold these investments. The more the largest banks diversify their investments, the more opportunities for shared contagion. Several studies have found that during a financial crisis, diversification can destabilise the wider network.[94]

Robert May and Andy Haldane noted that historically, the largest banks had held lower amounts of capital than their smaller peers. The popular argument was that because these banks held more diverse investments, they were at less risk; they didn't need to have a big buffer against unexpected losses. The 2008 crisis revealed the flaws in this thinking. Large banks were no less likely to fail than smaller ones. What's more, these big firms were disproportionally important to the stability of the financial network. 'What matters is not a bank's closeness to the edge of the cliff,' May and Haldane wrote in 2011, 'it is the extent of the fall.'[95]

TWO DAYS AFTER LEHMAN went under, *Financial Times* journalist John Authers visited a Manhattan branch of Citibank during his lunch break. He wanted to move some cash out of his account. Some of his money was covered by government deposit insurance, but only up to a limit; if Citibank collapsed too, he'd lose the rest. He wasn't the only one who'd had this idea. 'At Citi, I found a long queue, all well-dressed Wall Streeters,' he later wrote.[96] 'They were doing the same as me.' The bank staff helped him open additional accounts in the name of his wife and children, reducing his risk. Authers was shocked to discover they'd been doing this all morning. 'I was finding it a little hard to breathe. There was a bank run happening, in New York's financial district. The people panicking were the Wall Streeters who best understood what was going on.' Should he report what was happening? Given the severity of the crisis, Authers decided it would only make the situation worse. 'Such a story on the FT's front page might have been enough to push the system over the edge.' His counterparts at other newspapers came to the same conclusion, and the news went uncovered.

The analogy between financial and biological contagion is a useful starting point, but there is one situation it doesn't cover. To get infected during a disease outbreak, a person needs to be exposed to the pathogen. Financial contagion can also spread through tangible exposures, like a loan between banks or an investment in the same asset as someone else. The difference with finance is that firms don't always need a direct exposure to fall ill. 'There's one way this is unlike any other network we've dealt with,' said Nim Arinaminpathy. 'You can have apparently healthy institutions come crashing down.' If the public believes that a bank will go under, they may try to withdraw their money all at once, which would sink even a healthy bank. Likewise, when banks lose confidence in the financial system – as happened in 2007/8 – they often hoard money rather than lending it out. The rumour and speculation that circulates from one trader to another may therefore bring down firms that would otherwise have survived the crisis.

During 2011, Arinaminpathy and Robert May worked with Sujit Kapadia at the Bank of England to investigate not only direct transmission through bad loans or shared investments, but also the indirect effect of fear and panic. They found that if bankers started hoarding money when they lost confidence in the system, it could exacerbate a crisis: banks that would otherwise have had enough capital to ride it out would instead fail. The damage was much worse when a large bank was involved because they tended to be in the middle of the financial network.[97] This suggested that rather than simply looking at the size of banks, regulators should consider who is at the heart of the system. It isn't just about banks being 'too big to fail'; it is more about them being 'too central to fail'.

These kinds of insights from epidemic theory are now being put into practice, something Haldane described as a 'philosophical shift' in how we think about financial contagion. One major change has been to get banks to hold more capital if they are important to the network, reducing their susceptibility to infection. Then there is the issue of the network links that transmitted the infection in the first place. Could regulators target these too? 'The hardest part of this was when you went to questions of "Should we act to alter the very structure of the web"?' Haldane said. 'That's when people started to kick up more of a fuss because it was a more intrusive intervention in their business model.'

In 2011, a commission chaired by John Vickers recommended that larger British banks put a 'ring-fence' around their riskier trading activities.[98] This would help prevent the fallout from bad investments spreading to the retail parts of banks, which deal with high-street services like our savings accounts. 'The ring-fence would help insulate UK retail banking from external shocks,' the commission suggested. 'A channel of financial system interconnectedness – and hence of contagion – would be made safer.' The UK government eventually put the recommendation into practice, forcing banks to split their activities. Because it was such a tough policy to get through, it wasn't picked up elsewhere; ring-fencing was proposed in other parts of Europe, but not implemented.[99]

Ring-fencing isn't the only strategy for reducing transmission. When banks trade financial derivatives, it's often done 'over the counter' from one firm direct to another, rather than through a central exchange. Such trading activity came to almost $600 trillion in 2018.[100] However, since 2009, the largest derivatives contracts are no longer traded directly between major banks. They now have to go through independently run central hubs which have the effect of simplifying the network structure.

The danger, of course, is that if a hub fails, it could become a giant superspreader. 'If there is a big shock, it makes things worse because the risk is concentrated,' said Barbara Casu, an economist at Cass Business School.[101] 'It should act as a risk buffer, but in extreme cases it could act as a risk amplifier.' To guard against this problem, hubs have access to emergency capital from the members who use them. This mutual approach has drawn criticism from financiers who prefer an every-firm-for-themselves style of banking.[102] But by removing the tangle of hidden loops from the network, the hubs should mean fewer opportunities for contagion, and less uncertainty about who is at risk.

Despite progress in our understanding of financial contagion, there is still work to be done. 'It's like infectious disease modelling in the 1970s and 1980s,' said Arinaminpathy. 'There was a lot of great theory and the data had some catching up to do.' One of the big obstacles is access to trading information. Banks are naturally protective of their business activities, making it difficult for researchers to form a picture of exactly how institutions are connected, particularly at the global level. This makes it difficult to assess potential contagion. Network scientists have found that, when examining the probability of a crisis, small errors in knowledge about the lending network could lead to big errors in estimates of system-wide risk.[103]

Yet it's not only a matter of trading data. As well as studying the structure of networks, we need to think more about Newton's 'madness of people'. We need to consider how beliefs and behaviours arise, and how they can spread. This means thinking about people as well as pathogens. From innovations to infections, contagion is often a social process.

The measure of friendship

THE TERMS OF THE WAGER were simple. If John Ellis lost at darts, he had to get the word 'penguin' into his next scientific paper. It was 1977, and Ellis and his colleagues were in a pub near the CERN particle physics laboratory, just outside Geneva. Ellis was playing against Melissa Franklin, a visiting student. She had to leave before the end of the game, but another researcher took her place and sealed the victory. 'Nevertheless,' Ellis later said,[1] 'I felt obligated to carry out the conditions of the bet.'

That raised the question of how to sneak a penguin into a physics paper. At the time, Ellis was working on a manuscript that described how a particular type of subatomic particle – the so-called 'bottom quark' – behaved. As was common in physics, he sketched out a diagram with arrows and loops showing how the particles would transition from one state to another. First introduced by Richard Feynman in 1948, these 'Feynman diagrams' had become a popular tool for physicists. The drawings provided Ellis with the inspiration he needed. 'One evening, after working at CERN, I stopped on my way back to my apartment to visit some friends living in Meyrin where I smoked some illegal substance,' he recalled. 'Later, when I got back to my apartment and continued working on our paper, I had a sudden flash that the famous diagrams look like penguins.'

Ellis's idea would catch on. Since the paper was published, his 'penguin diagrams' have been cited thousands of times by other physicists. Even so, the penguins are nowhere near as widespread as the figures they are based on. Feynman diagrams would spread rapidly after their 1948 debut, transforming physics. One of the reasons the idea sparked was the Institute for Advanced Study in Princeton, New Jersey. Its director was J. Robert Oppenheimer, who'd previously led the US effort to develop the atomic bomb. Oppenheimer called the institute his 'intellectual hotel', bringing in a series of junior researchers on two-year positions.[2] Young minds arrived from around the world, with Oppenheimer wanting to encourage the global flow of ideas. 'The best way to send information is to wrap it up in a person,' as he put it.

The spread of scientific concepts would inspire some of the first research into the transmission of ideas. During the early 1960s, US mathematician William Goffman suggested that the transfer of information between scientists worked much like an epidemic.[3] Just as diseases like malaria spread from person to person via mosquitoes, scientific research often passed from scientist to scientist via academic papers. From Darwin's theory of evolution to Newton's laws of motion and Freud's psycho-analytic movement, new concepts had spread to 'susceptible' scientists who came into contact with them.

Still, not everyone was susceptible to Feynman diagrams. One sceptic was Lev Landau at the Moscow Institute for Physical Problems. A highly respected physicist, Landau had clear ideas about how much he respected others; he was known to maintain a list rating his fellow researchers. Landau used an inverted scale from 0 to 5. A score of 0 indicated the greatest physicist – a position held only by Newton in the list – and 5 meant 'mundane'. Landau rated himself a 2.5, upgrading this to a 2 after he won the 1962 Nobel Prize.[4]

Although Landau rated Feynman as a 1, he wasn't impressed

by the diagrams, seeing them as a distraction from more important problems. Landau hosted a popular weekly seminar at the Moscow Institute. Twice, speakers tried to present Feynman diagrams; both times they were kicked off the podium before they could finish their talks. When a PhD student said he was planning to follow Feynman's lead, Landau accused him of 'fashion chasing'. Landau did eventually use the diagrams in a 1954 paper, but he outsourced the tricky analysis to two of his students. 'This is the first work where I could not carry out the calculations myself', he admitted to a colleague.[5]

What effect did people like Landau have on the spread of Feynman diagrams? In 2005, physicist Luís Bettencourt, historian David Kaiser and their colleagues decided to find out.[6] Kaiser had previously collected academic journals published around the world in the years after Feynman announced his idea. He then went through each journal page-by-page, looking for references to Feynman diagrams, and tallying up how many authors adopted the idea over time. When the team plotted the data, the number of authors using the diagrams followed the familiar S-shaped adoption curve, rising exponentially before eventually plateauing.

The next step was to quantify how contagious the idea had been. Although the diagrams had originated in the US, they had spread quickly when they arrived in Japan. Things were more sluggish in the USSR, with a slower uptake than the other two countries. This was consistent with historical accounts. Japanese universities had expanded rapidly during the post-war period, with a strong particle physics community. In contrast, the emerging Cold War – combined with the scepticism of researchers like Landau – had stifled the diagrams in the USSR.

With the data they had available, Bettencourt and colleagues could also estimate the reproduction number, R, of a Feynman diagram: for each physicist who adopted the idea, how many others did they eventually pass it on to? Their results suggested a lot: as an idea, it was highly contagious. Initially R was around 15 in the USA and potentially as high as 75 in Japan. It was one of the first times that researchers had tried to measure the reproduction number of an idea, putting a number on what had previously been a vague notion of contagiousness.

This raised the question of why the idea had been so catchy. Perhaps it was because physicists were interacting with each other frequently during this period? Not necessarily: the high value of R instead seemed to be because people kept spreading the idea for a long time once they'd adopted it. 'The spread of Feynman diagrams appears analogous to a very slowly spreading disease,' the researchers noted. Adoption was 'due primarily to the very long lifetime of the idea, rather than to abnormally high contact rates'.

Tracing citation networks doesn't just tell us how new ideas spread. We can also learn how they emerge. If high profile scientists dominate a field, it can hinder the growth of competing ideas. As a result, new theories may only gain traction once dominant scientists cede the limelight. As physicist Max Planck supposedly once said, 'science advances one funeral at a time.' Researchers at MIT have since tested this famous comment by analysing what happens after the premature deaths of elite scientists.[7] They found that competing groups would subsequently publish more papers – and pick up more citations – while collaborators of the 'star' researcher tended to fade in prominence.

Scientific papers aren't only relevant to scientists. Ed Catmull, co-founder of Pixar, has argued that publications are a useful way of building links with specialists outside their company.[8] 'Publishing may give away ideas, but it keeps us connected with the academic community,' he once wrote. 'This connection is worth far more than any ideas we may have revealed'. Pixar is known for encouraging 'small-world' encounters between different parts of a network. This has even influenced the design of their building, which has a large central atrium containing potential hubs for random interactions, like mailboxes and the cafeteria. 'Most buildings are designed for some functional purpose, but ours is structured to maximize inadvertent encounters,' as Catmull put it. The idea of social architecture has caught on elsewhere too. In 2016, the Francis Crick Institute opened in London. Europe's largest biomedical lab, it would become home to over 1,200 scientists in a £650 million building. According to its director Paul Nurse, the layout was designed to get people interacting by creating 'a bit of gentle anarchy'.[9]

Unexpected encounters can help spark innovation, but if companies remove too many office boundaries, it can have the opposite effect. When researchers at Harvard University used digital trackers to monitor employees at two major companies, they found that the introduction of open-plan offices reduced face-to-face interactions by around 70 per cent. People instead chose to communicate online, with e-mail use increasing by over 50 per cent. Increasing the openness of the offices had decreased the number of meaningful interactions, reducing overall productivity.[10]

For something to spread, susceptible and infectious people need to come into contact, either directly or indirectly. Whether we're looking at innovations or infections, the number of opportunities for transmission will depend on how often contacts occur. If we want to understand contagion, we therefore need to work out how we interact with one another. However, it's a task that turns out to be remarkably difficult.

'THATCHER HALTS SURVEY ON SEX,' announced the headline in *The Sunday Times*. It was September 1989, and the government had just blocked a proposal to study sexual behaviour in the UK. Faced with a growing HIV epidemic, researchers had become increasingly aware of the importance of sexual encounters. The problem was that nobody really knew how common these encounters were. 'We had no idea of the parameter estimates that would drive an epidemic of HIV,' Anne Johnson, one of the researchers who'd proposed the UK study, later said. 'We didn't know what proportion of the population had gay partners, we didn't know the number of partners that people had.'[11]

In the mid-1980s, a group of health researchers had come up with the idea of measuring sexual behaviour on a national scale. They'd run a successful pilot study, but had struggled to get the main survey off the ground. There were reports that Margaret Thatcher had vetoed government funding, believing that the study would intrude into people's private lives, leading to 'unseemly speculation'. Fortunately, there was another option. Shortly after *The Sunday Times* article came out, the team secured independent support from the Wellcome Trust.

The National Survey of Sexual Attitudes and Lifestyles – or Natsal – would eventually run in 1990, then again in 2000 and 2010. According to Kaye Wellings, who helped develop the study, it was clear the data would have applications beyond STIs. 'Even as we were writing the proposal, I think we realised that it was going to answer a whole host of questions of relevance to public health policy, which there hadn't been data available to answer before.' In recent years, Natsal has provided insights into a whole range of social issues, from birth control to marriage breakdowns.

Still, it wasn't easy to get people talking about their sex lives. Interviewers had to persuade people to take part – often by emphasising the benefits for wider society – and build enough trust for participants to answer honestly. Then there was the issue of sexual terminology. 'There was that mismatch between the public health language and the language of everyday, which was so full of euphemisms,' Wellings noted. She recalled that several participants didn't recognise terms like 'heterosexual' or 'vaginal'. 'All the Latin-sounding names, or any word with more than three syllables, was thought of as something completely weird and unorthodox.'

Yet the Natsal team did have some advantages, such as the relatively low frequency of sexual encounters. The most recent Natsal study found that a typical twenty-something in the UK has sex about five times a month on average, with less than one new sexual partner per year.[12] Even the most active individuals are unlikely to sleep with more than a few dozen people in a given year. It means that most interviewees will know how many partners they've had and what those partnerships involved. Contrast that with the sort of interactions that might spread flu, such as conversations or handshakes. Each day, we may have dozens of face-to-face encounters like these.

During the past decade or so, researchers have increasingly tried to measure social contacts that are relevant for respiratory infections like flu. The best known is the POLYMOD study, which asked over 7,000 participants in eight European countries who they interacted with. This included physical contacts, like handshakes, as well as conversations. Researchers have since run similar studies in countries ranging from Kenya to Hong Kong. The studies are also getting more ambitious: in 2017/18 I worked with collaborators at the University of Cambridge to run a public science project collecting social behaviour data from over 50,000 volunteers in the UK.[13]

Thanks to these studies, we now know that certain aspects of behaviour are fairly consistent around the world. People tend to mix with people of a similar age, with children having by far the most contacts.[14] Interactions in schools and at home typically involve physical contact, and encounters that occur on a daily basis often last longer than an hour. Even so, the overall number of interactions can vary a lot between locations. Hong Kong residents typically have physical contact with around five other people each day; the UK is similar, but in Italy, the average is ten.[15]

It's one thing to measure such behaviour, but can this new information help predict the shape of epidemics? At the start

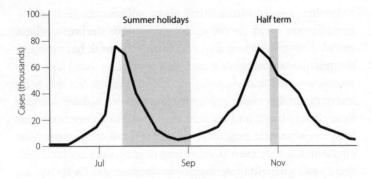

Dynamics of the 2009 influenza pandemic in the UK

of this book, we saw that during the 2009 influenza pandemic, there were two outbreak peaks in the United Kingdom: one in the spring and one in the autumn. To understand what caused this pattern, we simply need to look at schools. These bring children together in an intensely social environment, creating a potential mixing pot of infection; during the school holidays, children have around 40 per cent fewer daily social contacts on average. As you can see from the graph above, the gap between the two pandemic peaks in 2009 coincided with the school holiday. This lengthy drop in social contacts was large enough to explain the summer lull in the pandemic. However, school holidays can't fully explain the second wave of infection. Although the first peak was probably due to changes in social behaviour, the second peak was mostly down to herd immunity.[16]

It's been estimated that the 2009 epidemic would have been 20 per cent larger had it not been for this summer interruption in transmission.[17] The difference illustrates a crucial quirk of herd immunity. Without control measures or changes in behaviour, epidemics typically 'overshoot' and infect more people than required to reduce susceptibility to the point where the reproduction number is below one. Think about the shape of a simple epidemic with a single peak: by definition, R is below one after the peak (because the epidemic is declining). That means all the infections that occur after the peak happen after the population technically has herd immunity, but before the overall level infection has declined to low levels. If there is an interruption to transmission – such as school holiday – it can slow the epidemic and reduces the amount of overshoot that happens after the population reaches herd immunity. This effect means that even if an epidemic isn't fully controllable, there's still a benefit to slowing down transmission with control measures, because herd immunity will be reached with fewer infections overall.

The rise and fall of infections during school terms and holidays can influence other health conditions too. In many countries, asthma cases peak at the start of a school term. These outbreaks can also have a knock-on effect in the wider community, exacerbating asthma in adults.[18]

If we want to predict a person's risk of infection, it's not enough to measure how many contacts they have. We also need to think about their contacts' contacts, and *their* contacts' contacts. A person with seemingly few interactions might be just a couple of steps away from a high transmission environment like a school. A few years ago, my colleagues and I looked at social contacts and infections during the 2009 flu pandemic in Hong Kong.[19] We found that it was the high number of social contacts among children that drove the pandemic, with a drop in contacts and infection after childhood. But there was a subsequent increase in risk when people reached parenthood age. As any teacher or parent will know, interactions with children means an increased risk of infection. In the US, people without children in their house typically spend a few weeks of the year infected with viruses; people with one child have an infection for about a third of the year; and those with two children will on average carry viruses more often than not.[20]

As well as driving transmission in communities, social inter-actions can also transport infections to other locations. In the early stages of the 2009 flu pandemic, the virus didn't spread according to the as-the-crow-flies distance between countries. When the outbreak started in Mexico in March, it quickly reached faraway places like China, but took longer to appear in nearby countries such as Barbados. The reason? If we define 'near' and 'far' in terms of locations on a map, we're using the wrong notion of distance. Infections are spread by people, and there are more major flight routes linking Mexico and China – such as those via London – than those connecting Mexico with places like Barbados. China might be far away for a crow, but it's relatively close for a human. It turns out that the spread of flu in 2009 is much easier to explain if we instead define distances according to airline passenger flows. And not just flu: SARS followed similar airline routes when it emerged in China in 2003, arriving in countries like the Republic of Ireland and Canada before Thailand and South Korea.[21]

Once the 2009 flu pandemic arrived in a country, however, long travel distance seemed to be less important for transmission. In the US, the virus spread like a ripple, gradually travelling from the southeast outwards. It took about three months to move 2,000 kilometres across the eastern US, which works out at a speed of just under 1 km/h. On average, you could have outwalked it.[22]

Although long-distance flight connections are important for introducing viruses to new countries, travel within the US is dominated by local movements. The same is true of many other countries.[23] To simulate these local movements, research-ers often use what's known as a 'gravity model'. The idea is that we are drawn to places depending on how close and popu-lous they are, much like larger, denser planets have a stronger gravitational pull. If you live in a village, you might visit a nearby town more often than a city further away; if you live in a city, you'll probably spend little time in the surrounding towns.

This might seem like an obvious way to think about inter-actions and movements, but historically people have thought otherwise. In the mid-1840s, at the peak of Britain's railway bubble, engineers assumed that most traffic would come from long-distance travel between big cities. Unfortunately, few both-ered to question this assumption. There were some studies on the continent, though. To work out how people might actu-ally travel, Belgian engineer Henri-Guillaume Desart designed the first ever gravity model in 1846. His analysis showed that there would be a lot of demand for local trips, an idea that was ignored by rail operators on the other side of the channel. The British railway network would probably have been far more effi-cient had it not been for this oversight.[24]

It can be easy to underestimate the importance of social ties. When Ronald Ross and Hilda Hudson wrote those papers on the 'theory of happenings' in the early twentieth century, they suggested it could apply to things like accidents, divorce and chronic diseases. In their minds, these things were independ-ent happenings: if something happened to one person, it didn't affect the chances of it happening to someone else. There was no element of contagion from one person to another. At the start of the twenty-first century, researchers started to question whether this was really the case. In 2007, physician Nicholas Christakis and social scientist James Fowler published a paper titled 'The Spread of Obesity in a Large Social Network over 32 Years'. They had studied health data from participants in the long-running Framingham Heart Study, based in the city of Framingham, Massachusetts. As well as suggesting that obesity could spread between friends, they proposed that there could be a knock-on effect further into the network, potentially influenc-ing friends-of-friends and friends-of-friends-of-friends.

The pair subsequently looked at several other forms of social contagion in the same network, including smoking, happiness, divorce, and loneliness.[25] It might seem odd that loneliness could spread through social contacts, but the researchers pointed to what might be happening at the edge of a friendship network. 'On the periphery, people have fewer friends, which makes them lonely, but it also drives them to cut the few ties that they have left. But before they do, they tend to transmit the same feeling of loneliness to their remaining friends, starting the cycle anew.'

These papers have been hugely influential. In the decade after it was published, the obesity study alone was cited over 4,000 times, with many seeing the research as evidence that such traits can spread. But it's also come under fire. Soon after the obesity and smoking studies were published, a paper in the *British Medical Journal* suggested that Christakis and Fowler's analysis might have flagged up effects that weren't really there.[26] Then mathematician Russell Lyons wrote a paper arguing that the researchers had made 'fundamental errors' and that 'their major claims are unfounded'.[27] So where does that leave us? Do things like obesity actually spread? How do we even work out if behaviour is contagious?

ONE OF THE MOST FAMILIAR EXAMPLES of social contagion is yawning, and it's also one of the easiest forms of contagion to study. Because it's common, easy to spot, and the delay from one person's yawn to another is relatively short, researchers can look at transmission in detail.

By setting up lab experiments, several studies have analysed what makes yawns spread. The nature of social relationships seems to be particularly important for transmission: the better we know someone, the more likely it is that we'll catch their yawn.[28] The transmission process is also faster, with a smaller delay between yawns among family members than among acquaintances. Yawn in front of a stranger and there's a less than 10 per cent chance it will spread; yawn near a family member and they'll catch it in about half the time. It's not just humans who are more likely to pick up yawns from individuals they care about. Similar social yawning can occur among animals, from monkeys to wolves.[29] However, it can take a while for us to become susceptible to a yawn. Although infants and toddlers sometimes yawn, they don't seem to catch them from their parents. Experiments suggest yawning doesn't become contagious until children reach about four years old.[30]

As well as yawning, researchers have looked at the spread of other short-term behaviours, like itching, laughter, and emotional reactions. These social responses can manifest on very fast timescales: in experiments looking at teamwork, leaders were able to spread a positive or negative mood to their team in a matter of minutes.[31]

If researchers want to study yawning or mood, they can use laboratory set-ups to control what people see, and avoid distractions that could skew results. This is feasible for things that spread quickly, but what about behaviours and ideas that take much longer to propagate through a population? It's much harder to study social contagion outside a laboratory. This isn't just a challenge for human populations. Among birds, great tits have a long-standing reputation for innovation. In the 1940s, British ecologists noted that they had worked out how to peck through the foil of milk bottles to get at the cream. The tactic would persist for decades, but it wasn't clear how such innovations spread through bird populations.[32]

Although several studies have looked at the spread of animal behaviour in captivity, it has been difficult to do the same in wild populations. Given great tits' reputation for innovation, zoologist Lucy Aplin and her colleagues set out to see how these ideas propagated. First they needed a new innovation. The team headed out into Wytham Woods, near Oxford, and set up a puzzle box containing mealworms. If the birds wanted to get the food inside, they'd need to move a sliding door in a certain direction. To see how the birds interacted, the researchers tagged almost all the tits in the area with automated tracking devices. 'We could get real-time information about how and when individuals acquired knowledge,' Aplin said. 'The automated data-collection also meant we could let the process run without disturbance.'[33]

The birds grouped together into several different sub-populations; in five of these populations, the researchers taught a couple of birds how to solve the puzzle. The technique spread quickly: within twenty days, three in every four birds had picked up the idea. The team also studied a control group of birds, which hadn't been trained. A few eventually worked out how to get into the box, but it took much longer for the idea to emerge and spread.

In the trained populations, the idea was also highly resilient. Many of the birds died from one season to the next, but the knowledge didn't. 'The behaviour re-emerged very quickly each winter,' Aplin said, 'even if there were only a small number of individuals that were alive from the previous year and had knowledge of the behaviour.' She also noticed that transmission of information between birds had some familiar features. 'Some general principles are similar to how disease spreads through populations, for instance more social individuals being more likely to encounter and adopt new behaviours, and socially central individuals can act as "keystones" or "super-spreaders" in the diffusion of information.'

The study also demonstrated that social norms could emerge in wild animals. There were actually a couple of ways to get into the puzzle box, but it was the solution the researchers had introduced that became the accepted method. Such conformity is even more common when we look at humans. 'We're social learning specialists,' Aplin said. 'The social learning and culture

we observe in human societies is of a magnitude greater than
anything we observe in the rest of the animal kingdom.'

WE OFTEN SHARE CHARACTERISTICS with people we know, from
health and lifestyle choices to political views and wealth. In
general, there are three possible explanations for such similarities. One is social contagion: perhaps you behave in a certain
way because your friends have influenced you over time. Alternatively, it may be the other way around: you may have chosen
to become friends because you already shared certain characteristics. This is known as 'homophily', the idea that 'birds
of a feather flock together'. Of course, your behaviour might
be nothing to do with social connections at all. You may just
happen to share the same environment, which influences your
behaviour. Sociologist Max Weber used the example of a crowd
of people opening umbrellas when it starts to rain. They aren't
necessarily reacting to each other; they're reacting to the clouds
above.[34]

It can be tough to work out which of the three explanations
– social contagion, homophily or a shared environment – is the
correct one. Do you like a certain activity because your friend
does, or are you friends because you both like that activity? Did
you skip your running session because your friend did, or did
you both abandon the idea because it was raining? Sociologists
call it 'the reflection problem', because one explanation can
mirror another.[35] Our friendships and behaviour will often be
correlated, but it can be very difficult to show that contagion is
responsible.

What we need is a way to separate social contagion from the
other possible explanations. The most definitive way to do this
would be to spark an outbreak and watch what happens. This
would mean introducing a specific behaviour, like Aplin and her
colleagues did with birds, and measuring how it spreads. Ideally

we would compare results with a randomly selected 'control' group of individuals – who aren't exposed to the spark – to see how much effect the outbreak has. This type of experiment is common in medicine, where it's known as a 'randomised controlled trial'.

How might such an approach work in humans? Say we wanted to run an experiment to study the spread of cigarette smoking between friends. One option would be to introduce the behaviour we're interested in: pick some people at random, get them to take up smoking, and then see whether the behaviour spreads through their friendship groups. Although this experiment might tell us whether social contagion occurs, it doesn't take much to spot that there are some big ethical problems with this approach. We can't ask people to adopt a harmful activity like smoking on the off chance it will help us understand social behaviour.

Rather than randomly introducing smoking, we could instead look at how existing smoking behaviour spreads through new social connections. But this would mean rearranging people's friendships and locations at random and tracking whether people adopt their new friends' behaviour. Again, this is generally not feasible: who wants to reshuffle their entire friendship network for a research project?

When it comes to designing social experiments, Aplin's work on birds had some big advantages over studies of humans. Whereas humans may keep similar social links for years or decades, birds have a relatively short lifespan, which meant new networks of interactions would form each year. The team could also tag most of the birds in the area, making it possible to track the network in real-time. This meant the researchers could introduce a new idea – the puzzle solution – and watch how it spread through the newly formed networks.

There are some circumstances in which new human friendships randomly form all at once, for example when recruits are assigned to military squadrons or students are allocated to university halls.[36] Unfortunately for researchers, these are rare examples. In most real-life situations, scientists can't meddle with behaviour or friendship dynamics to see what might happen. Instead, they must try and gain insights from what they can observe naturally. 'Though a lot of the best strategies involve randomisation or some plausible source of randomness, for many things we really care about as social scientists and citizens, we're not going to be able to randomise,' said Dean Eckles, a social scientist at MIT.[37] 'So we should do the best job we can with purely observational research.'

Much of epidemiology relies on observational analysis: in general, reseachers can't deliberately start outbreaks or give people severe illnesses to understand how they work. This has led to some suggestions that epidemiology is closer to journalism than science, because it just reports on the situation as it happens, instead of running experiments.[38] But such claims ignore the huge improvements in health that have come from observational studies.

Take smoking. In the 1950s, researchers started to investigate the massive rise in lung cancer deaths that had occurred during the preceding decades.[39] There seemed to be a clear link with the popularity of cigarettes: people who smoked were nine times more likely to die of the disease than non-smokers. The problem was how to show that smoking was actually causing cancer. Ronald Fisher, a prominent statistician (and heavy pipe smoker) argued that just because the two things were correlated, it didn't mean one was causing the other. Perhaps smokers had very different lifestyles to non-smokers, and it was one of these differences, rather than smoking, that was causing the deaths? Or maybe there was some genetic trait – as yet unidentified – that happened to make people both more likely to develop lung cancer and more likely to smoke? The issue divided the scientific community. Some, like Fisher, argued that the patterns linking

smoking and cancer were just a coincidence. Others, like epidemiologist Austin Bradford Hill, thought that smoking was to
blame for the rising deaths.

Of course, there was an experiment that would have given
a definitive answer, but as we've already seen, it wouldn't have
been ethical to run it. Just as modern social scientists can't make
people take up smoking to see if the habit spreads, researchers
in the 1950s couldn't ask people to smoke to find out if it caused
cancer. To solve the puzzle, epidemiologists had to find a way
to work out whether one thing causes another without running
an experiment.

RONALD ROSS SPENT AUGUST 1898 waiting to announce his discovery that mosquitoes transmitted malaria. While he battled to
get government permission to publish the work in a scientific
journal, he feared others would pounce on his research and take
the credit. 'Pirates lay in the offing ready to board me,' as he
put it.[40]

The pirate he feared most was a German biologist named
Robert Koch. Stories were circulating that Koch had travelled
to Italy to study malaria. If he managed to infect a person with
the parasite, it could overshadow Ross's work, which had used
only birds. Relief came a few weeks later, in the form of a letter
from Patrick Manson. 'I hear Koch has failed with the mosquito
in Italy,' Manson wrote, 'so you have time to grab the discovery
for England.'

Eventually Koch did publish a series of malaria studies,
which fully credited Ross's work. In particular, Koch suggested
that children in malarial areas acted as reservoirs of infection,
because older adults had often developed immunity to the
parasite. Malaria was the latest in a line of new pathogens for
Koch. During the 1870s and 1880s, he had shown that bacteria
were behind diseases like anthrax in cattle and tuberculosis in

humans. In the process, he'd come up with a set of rules – or 'postulates' – to identify whether a particular germ is responsible for a disease. To start with, he thought that it should always be possible to find the germ inside someone who has the disease. Then, if a healthy host – like a laboratory animal – was exposed to this germ, it should develop the disease too. Finally, it should be possible to extract a sample of the germ from the new host once they fall ill; this germ should be the same as the one they were originally exposed to.[41]

Koch's postulates were useful for the emerging science of 'germ theory', but he soon realised they had limitations. The biggest problem was that some pathogens don't always cause disease. Sometimes people would get infected but not have noticeable symptoms. Researchers therefore needed a more general set of principles to work out what might be behind a disease.

For Austin Bradford Hill, the disease of interest was lung cancer. To show that smoking was responsible, he and his collaborators would eventually compile several types of evidence. He'd later summarise these as a set of 'viewpoints', which he hoped would help researchers decide whether one thing causes another. First on his list was the strength of correlation between the proposed cause and effect. For example, smokers were much more likely to get lung cancer than non-smokers. Bradford Hill said this pattern should be consistent, cropping up in different places across multiple studies. Then there was timing: did the cause come before the effect? Another indicator was whether the disease was specific to a certain type of behaviour (although this isn't always helpful because non-smokers can get lung cancer too). Ideally there would also be evidence from an experiment: if people stopped smoking, it should reduce their chances of cancer.

In some cases, Bradford Hill said it's possible to relate the level of exposure to the risk of disease. For instance, the more cigarettes a person smokes, the more likely they are to die from them. What's more, it may be possible to draw an analogy with a similar cause and effect, such as another chemical that causes cancer. Finally, Bradford Hill suggested it's worth checking to see whether the cause is biologically plausible and fits with what's already known to scientists.

Bradford Hill emphasised that these viewpoints were not a checklist to 'prove' something beyond dispute. Rather, the aim was to help answer a crucial question: is there any better explanation for what we are seeing than simple cause and effect? As well as providing evidence that smoking caused cancer, these kinds of methods have helped researchers uncover the source of other diseases. During the 1950s and 1960s, epidemiologist Alice Stewart gathered evidence that low-dose radiation could cause leukaemia.[42] At the time, new X-ray technology was regularly being used on pregnant women; there were even X-rays in shoe shops, so people could see their feet inside the shoes. After a long battle by Stewart, these hazards were removed. More recently, researchers at the US CDC used the Bradford Hill viewpoints to argue that infections with Zika were causing birth defects.[43]

Establishing such causes and effects is inherently difficult. Often there will be an intense debate about what is responsible and what should be done. Still, Stewart believed that, faced with troubling evidence, people should act despite the inevitable uncertainty involved. 'The trick is to get the best guess of the thickness of the ice when crossing a lake,' she once said. 'The art of the game is to get the correct judgment of the weight of the evidence, knowing that your judgment is subject to change under the pressure of new observations.'[44]

WHEN CHRISTAKIS AND FOWLER originally set out to study social contagion, they'd planned to do it from scratch. The idea was to recruit 1,000 people, get each of them to name five contacts, and then get each of their contacts to name five more contacts. In total, they would have had to track the behaviour of 31,000 people in detail for multiple years. A study that large would have cost around $30m.[45]

While exploring options, the pair got in touch with the team running the Framingham Heart Study, because it would be easier to recruit those initial 1,000 people from an existing project. When Christakis visited Marian Bellwood, the project co-ordinator, she mentioned they kept forms in the basement with details of each participant. To avoid losing contact with participants, they'd got people to list their relatives, friends and co-workers on the forms. It turned out that many of these contacts were also in the study, which meant their health information was being recorded too.

Christakis was astonished. Rather than recruiting a completely new set of social contacts, they could instead piece together the social network among Framingham participants. 'I called James from the parking lot and said, "you won't believe this!",' he recalled. There was just one catch: they'd have to go through twelve thousand names and fifty thousand addresses to identify the existing links. 'We had to decipher everyone's handwriting,' Christakis said. 'It took two years to computerise it.'

The pair had initially thought about analysing the spread of smoking, but decided obesity was a better starting point. Smoking depended on what participants reported, whereas obesity could be observed directly. 'Because we were doing something so novel, we wanted to start with something that could be objectively measured,' Christakis said.

The next step was to estimate whether obesity was being transmitted through the network. This meant tackling the reflection problem, separating potential contagion from homophily or environmental factors. To try and rule out the birds-of-a-feather effect of homophily, the pair included a time lag in the analysis; if obesity really spread from one person to their friend, the friend couldn't have become obese first. Environmental factors were trickier to exclude, but Christakis and Fowler tried to tackle the issue by looking at the direction of friendship. Suppose I list you as a friend in a survey, but you don't list me. This suggests I am more influenced by you than you are by me. However, if in reality we're actually both influenced by some shared environmental factor – like a new fast food restaurant – our friendship direction shouldn't affect who becomes obese. Christakis and Fowler found evidence that it did matter, suggesting that obesity could be contagious.

When the analysis was published, it received sharp criticism from some researchers. Much of the debate came down to two main points. The first was that the statistical evidence could have been stronger: the result showing that obesity was contagious was not as definitive as it would need to be for, say, a clinical trial showing whether a new drug worked. The second criticism was that, given the methods and data Christakis and Fowler had used, they could not conclusively rule out other explanations. In theory, it was possible to imagine a situation involving homophily and environment that could have produced the same pattern.

In my view, these are both reasonable criticisms of the research. But it doesn't mean that the studies weren't useful. Commenting on the debate about Christakis and Fowler's early papers, statistician Tom Snijders suggested that the studies had limitations, but were still important because they'd found an innovative way to put social contagion on scientists' agenda. 'Bravo for the imagination and braveness of Nick Christakis and James Fowler.'[46]

In the decade since Christakis and Fowler published their initial analysis of the Framingham data, evidence for social contagion has accumulated. Several other research groups have also shown that things like obesity, smoking, and happiness can be contagious. As we've seen, it is notoriously difficult to study social contagion, but we now have a much better understanding of what can spread.

The next step will be to move beyond simply saying that contagion exists. Showing that behaviour can catch on is equivalent to knowing that the reproduction number is above zero: on average, there will be some transmission, but we don't know how much. Of course, this is still useful information, because it shows contagion is a factor we need to think about. It tells us the behaviour is capable of spreading, even if we can't predict how big the outbreak might be. However, if governments and other organisations want to address health issues that are contagious, they'll need to know more about the actual extent of social contagion, and what impact different policies might have. If one person in a friendship group becomes overweight, exactly how much influence will it have on others? If you become happier, how much will your community's happiness increase? Christakis and Fowler have acknowledged that it's tricky to estimate the precise extent of social contagion. What's more, addressing such questions often means using imperfect data and methods. But as new datasets become available, they point out others will be able to build on their analysis, moving towards an accurate measurement of contagion.

By studying potentially contagious behaviour, researchers are also uncovering some crucial differences between biological and social outbreaks. In the 1970s, sociologist Mark Granovetter suggested that information could spread further through acquaintances than through close friends. This was because friends would often have multiple links in common, making most transmission redundant. 'If one tells a rumor to all his close friends, and they do likewise, many will hear the rumor a second and third time, since those linked by strong ties tend to share friends.' He referred to the importance of acquaintances as the 'strength of weak ties': if you want access to new information, you may be more likely to get it through a casual contact than a close friend.[47]

These long distance links have become a central part of network science. As we've seen, 'small-world' connections can help biological and financial contagion jump from one part of a network to another. In some cases, these links may also save lives. There is a long-standing paradox in medicine: people who have a heart attack or stroke while surrounded by relatives take longer to get medical care. This may well be down to the structure of social networks. There's evidence that close-knit groups of relatives tend to prefer a wait-and-see approach after witnessing a mild stroke, with nobody willing to contradict the dominant view. In contrast, 'weak ties' – like co-workers or non-relatives – can bring a more diverse set of perspectives, so flag up symptoms faster and call for help sooner.[48]

Even so, the sort of network structure that amplifies disease transmission won't always have the same effect on social contagion. Sociologist Damon Centola points to the example of HIV, which has spread widely through networks of sexual partners. If biological and social contagion work in the same way, ideas about preventing the disease should also have spread widely via these networks. And yet they have not. Something must be slowing the information down.

During an infectious disease outbreak, infection typically spreads through a series of single encounters. If you get the infection, it will usually have come from a specific person.[49] Things aren't always so simple for social behaviour. We might only start doing something after we've seen multiple other people doing it, in which case there is no single clear route of transmission. These behaviours are known as 'complex contagions', because transmission requires multiple exposures. For example, in Christakis and Fowler's analysis of smoking, they noted that people were more likely to quit if lots of their contacts stopped as well. Researchers have also identified complex contagion in behaviours ranging from exercise and health habits to the uptake of innovations and political activism. Whereas a pathogen like HIV can spread through a single long-range contact, complex contagions need multiple people to transmit them, so can't pass through single links. While small-world networks might help diseases spread, these same networks could limit the transmission of complex contagions.

Why do complex contagions occur? Damon Centola and his colleague Michael Macy have proposed four processes that might explain what's happening. First, there can be benefits to joining something that has existing participants. From social networks to protests, new ideas are often more appealing if more people have already adopted them. Second, multiple exposures can generate credibility: people are more likely to believe in something if they get confirmation from several sources. Third, ideas can depend on social legitimacy: knowing about something isn't the same as seeing others acting – or not acting – on it. Take fire alarms. As well as signaling there might be a fire, alarms make it acceptable for everyone to leave the building. One classic 1968 experiment had students sit working in a room as it slowly filled with fake smoke.[50] If they were alone, they would generally respond; if they were with a group of studious actors, they would continue to work, waiting for someone else to react. Finally, we have the process of emotional amplification. People may be more likely to adopt certain ideas or behaviours amid the intensity of a social gathering: just think about the collective emotion that comes with something like a wedding or a music concert.

The existence of complex contagions means we may need to re-evaluate what makes innovations spread. Centola has suggested that intuitive approaches for making things catch on may not work so well if people need multiple prompts to adopt an idea. To get innovation to spread in business, for example, it's not enough to simply encourage more interactions within an organisation. For complex contagions to spread, interactions need to be clustered together in a way that allows social reinforcement of ideas; people may be more likely to adopt a new behaviour if they repeatedly see everyone in their team doing it. However, organisations can't be too cliquey, otherwise new ideas won't spread beyond a small group of people. There needs to be a balance in the network of interactions: as well as having local teams acting as incubators for ideas, there are benefits to having Pixar-style overlaps between groups to get innovations out to a wider audience.[51]

The science of social contagion has come a long way in the past decade, but there is still much more to discover. Not least because it's often difficult to establish whether something is contagious in the first place. In many cases, we can't deliberately change people's behaviour, so we have to rely on observational data, as Christakis and Fowler did with the Framingham study. However, there is another approach emerging. Researchers are increasingly turning to 'natural experiments' to examine social contagion.[52] Rather than imposing behavioural change, they instead wait for nature to do it for them. For example, a runner in Oregon might change their routine when the weather is bad; if their friend in California changes their behaviour too, it could suggest social contagion is responsible. When researchers at MIT looked at data from digital fitness trackers, which included a social network linking users, they found that the weather could indeed reveal patterns of contagion. However, some were more likely to catch the running bug than others. Over a five-year period, the behaviour of less active runners tended to influence more active runners, but not the other way around. This implies that keen runners don't want to be outdone by their less energetic friends.

Behavioural nudges like changes in weather are a useful tool for studying contagion, but they do have limits. A rainy day might alter someone's running patterns, but it's unlikely to affect other, more fundamental behaviours like their marital choices or political views. Dean Eckles points out there can be a big gap between what is easily changed and what we ideally want to study. 'A lot of the behaviours we care the most about are not so easy to nudge people to do.'

IN NOVEMBER 2008, Californians voted to ban same-sex marriage. The result came as a shock to those who'd campaigned for marriage equality, especially as pre-vote polls had appeared to be in their favour. Explanations and excuses soon began to emerge. Dave Fleischer, director of the Los Angeles LGBT Center, noticed that several misconceptions about the result were becoming popular. One was that the people who voted for the ban must have hated the LGBT community. Fleischer disagreed with this idea. 'The dictionary defines "hate" as extreme aversion or hostility,' he wrote after the vote. 'This does not describe most who voted against us.'[53]

To find out why so many people were against same-sex marriage, the LGBT Center spent the next few years conducting thousands of face-to-face interviews. Canvassers used most of this time to listen to voters, a method known as 'deep canvassing'.[54] They encouraged people to talk about their lives, and reflect on their own experiences of prejudice. As they conducted these interviews, the LGBT Center realised that deep canvassing wasn't just providing information; it appeared to be changing voters' attitudes. If so, this would make it a powerful canvassing method. But was it really as effective as it seemed?

If people are rational, we might expect them to update their beliefs when presented with new information. In scientific research this approach is known as 'Bayesian reasoning'. Named after eighteenth-century statistician Thomas Bayes, the idea is to treat knowledge as a belief that we have a certain level of confidence in. For example, suppose you are strongly considering marrying someone, having thought carefully about the relationship. In this situation, it would take a very good reason for you to change your mind. However, if you're not totally sure about the relationship, you might be persuaded against marriage more easily. Something that might seem trivial to the infatuated may be enough to tip a wavering mind towards a break-up. The same logic applies to other situations. If you start with a firm belief, you'll generally need strong evidence to overcome it; if you are unsure at first, it might not take much for you to change your opinion. Your belief after exposure to new information therefore depends on two things: the strength of your initial belief and the strength of the new evidence.[55] This concept is at the heart of Bayesian reasoning – and much of modern statistics.

Yet there are suggestions that people don't absorb information in this way, especially if it goes against their existing views. In 2008, political scientists Brendan Nyhan and Jason Reifler proposed that persuasion can suffer from a 'backfire effect'. They'd presented people with information that conflicted with their political ideology, such as the lack of weapons of mass destruction in Iraq before the 2003 war, or the decline in revenues following President Bush's tax cuts. But it didn't seem to convince many of them. Worse, some people appeared to become more confident in their existing beliefs after seeing the new information.[56] Similar effects had come up in other psychological studies over the years. Experiments had tried to persuade people of one thing, only for them to end up believing something else.[57]

If the backfire effect is common, it doesn't bode well for canvassers hoping to convince people to change their minds about issues like same-sex marriage. The Los Angeles LGBT Center thought they had a method that worked, but it needed to be evaluated properly. In early 2013, Dave Fleischer had lunch with Donald Green, a political scientist at Columbia University. Green introduced Fleischer to Michael LaCour, a graduate student at UCLA, who agreed to run a scientific study testing the effectiveness of deep canvassing. The aim was to carry out

a randomised controlled trial. After recruiting voters to participate in a series of surveys, LaCour would randomly split the
group. Some would get visits from a canvasser; others, acting
as a control group, would have conversations about recycling.

What happened next would reveal a lot about how beliefs
change, just not quite in the way we might expect. It started
when LaCour reported back with some remarkable findings.
His trial had shown that when interviewers used deep canvassing methods, there was a large increase in interviewees' support
for same-sex marriage on average. Even better, the idea often
stuck, with the new belief still there months later. This belief
was also contagious, spreading to interviewees' housemates.
LaCour and Green published the results in the journal *Science*
in December 2014, attracting widespread media attention. It
seemed to be a stunning piece of research, showing how a small
action could have a massive influence.[58]

Then a pair of graduate students at the University of Berkeley noticed something strange. David Broockman and Joshua
Kalla had wanted to run their own study, building on LaCour's
impressive analysis. 'The most important paper of the year. No
doubt,' Broockman had told a journalist after the *Science* paper
was published. But when they looked at LaCour's dataset, it
seemed far too pristine; it was almost as if someone had simulated the data rather than collecting it.[59] In May 2015, the pair
contacted Green with their concerns. When questioned, LaCour
denied making up the data, but couldn't produce the original
files. A few days later, Green – who said he'd been unaware of
the problems until that point – asked *Science* to retract the paper.
It wasn't clear exactly what had happened, but it was clear that
LaCour hadn't run the study he said he had. The scandal came
as a huge disappointment to the Los Angeles LGBT Center. 'It
felt like a big punch to our collective gut,' said Laura Gardiner,
one of their organisers, after the problems emerged.[60]

Media outlets quickly added corrections to their earlier stories, but perhaps journalists – and the scientific journal – should have been more sceptical in the first place. 'What interests me is the repeated insistence on how unexpected and unprecedented this result was,' wrote statistician Andrew Gelman after the paper was retracted. Gelman pointed out that this seems to happen a lot in psychological science. 'People argue simultaneously that a result is completely surprising and that it makes complete sense.'[61] Although the backfire effect had been widely cited as a major hurdle to persuasion, here was a study claiming it could be cleared in one short conversation.

The media has a strong appetite for concise yet counter-intuitive insights. This encourages researchers to publicise results that show how 'one simple idea' can explain everything. In some cases, the desire for surprising-yet-simple conclusions can lead apparent experts to contradict their own source of expertise. Antonio García Martínez, who spent two years working in Facebook's ads team, recalled such a situation in his book *Chaos Monkeys*. Martínez tells the story of a senior manager who built a reputation with pithy, memorable insights about social influence. Unfortunately for the manager, these claims were undermined by research from his company's own data science team, whose rigorous analysis had shown something different.

In reality, it's very difficult to find simple laws that apply in all situations. If we have a promising theory, we therefore need to seek out examples that don't fit. We need to work out where its limits are and what exceptions there might be, because even widely reported theories might not be as conclusive as they seem. Take the backfire effect. After reading about the idea, Thomas Wood and Ethan Porter, two graduate students at the University of Chicago, set out to see how common it might actually be. 'Were the backfire effect to be observed across a population, the implications for democracy would be dire,' they wrote.[62] Whereas Nyhan and Reifler had focused on three main misconceptions, Wood and Porter tested thirty-six beliefs across

8,100 participants. They found that although it can be tough to convince people they're wrong, an attempted correction doesn't necessarily make their existing belief stronger. In fact, only one correction backfired in the study: the false claim about weapons of mass destruction in Iraq. 'By and large, citizens heed factual information, even when such information challenges their partisan and ideological commitments,' they concluded.

Even in their original study, Nyhan and Reifler found that the backfire effect is not guaranteed. During the 2004 presidential campaign, Democrats claimed that George Bush had banned stem cell research, whereas in reality, he'd limited funding for certain aspects of it.[63] When Nyhan and Reifler corrected this belief among liberals, the information was often ignored, but didn't backfire. 'The backfire effect finding got a lot of attention because it was so surprising,' Nyhan later said.[64] 'Encouragingly, it seems to be quite rare.' Nyhan, Reifler, Wood and Porter have since teamed up to explore the topic further. For example, in 2019 they reported that providing fact-checks during Donald Trump's election speeches had changed people's beliefs about his specific claims, but not their overall opinion of the candidate.[65] It seems some aspects of people's political beliefs are harder to alter than others. 'We have a lot more to learn,' Nyhan said.

When examining beliefs, we also need to be careful about what we mean by a backfire. Nyhan has noted that there can be confusion between the backfire effect and a related psychological quirk known as 'disconfirmation bias'.[66] This is when we give more scrutiny to arguments that contradict our existing beliefs than those that we agree with. Whereas the backfire effect implies that people ignore opposing arguments and strengthen their existing beliefs, disconfirmation bias simply means they tend to ignore arguments they view as weak.

It might seem like a subtle difference, but it's a crucial one. If the backfire effect is common, it implies that we can't persuade people with conflicting opinions to change their stance. No matter how convincing our arguments, they will only retreat further into their beliefs. Debate becomes hopeless and evidence worthless. In contrast, if people suffer from disconfirmation bias, it means their views could change, given compelling enough arguments. This creates a more optimistic outlook. Persuading people may still be challenging, but it is worth trying.

A lot rides on how we structure and present our arguments. In 2013, the UK legalised same-sex marriage. John Randall, then a Conservative MP, voted against the bill, a decision he later said he regretted. He wished he'd talked to one of his friends in Parliament beforehand, someone who – to many's surprise – had voted in favour of marriage equality. 'He said to me that it was something that wouldn't affect him at all but would give great happiness to many people,' Randall recalled in 2017. 'That is an argument that I find it difficult to find fault with.'[67]

Unfortunately, there is a major obstacle when it comes to finding a persuasive argument. If we have a strong opinion, Bayesian reasoning implies that we will struggle to distinguish the effects of arguments that support this existing view. Suppose you strongly believe in something. It could be anything from a political stance to an opinion about a film. If someone presents you with evidence that is consistent with your belief – regardless of whether this evidence is compelling or weak – you will go away with a similar opinion afterwards. Now imagine someone makes an argument against your belief. If that argument is weak, you won't change your view, but if it is watertight, you might well do so. From a Bayesian point of view, we are generally better at judging the effect of arguments that we disagree with.[68]

That's if we even think about different arguments. A few years ago, social psychologists Matthew Feinberg and Robb Willer asked people to come up with arguments that would persuade someone with an opposing political view. They found that many people used arguments that matched their own moral position, rather than the position of the person they were trying to persuade. Liberals tried to appeal to values like equality and social justice, while conservatives based their argument on things like loyalty and respect for authority. Arguing on familiar ground might have been a common strategy, but it wasn't an effective one; people were far more persuasive when they tailored their argument to the moral values of their opponent. This suggests that if you want to persuade a conservative, you're better off focusing on ideas like patriotism and community, whereas a liberal will be more convinced by messages promoting fairness.[69]

Even if you manage to identify an effective argument to support your position, there are things you can do to improve your chances of persuasion. First, the delivery method can matter. There's evidence that people are much more likely to complete a survey if asked in person rather than by e-mail,[70] for example. Other experiments have come to similar conclusions, finding that people can be more convincing face-to-face than by phone, post or online.[71]

The timing of messages can also make a difference. According to Briony Swire-Thompson, a psychologist at Northeastern University, researchers are increasingly thinking about how ideas wane. 'It's this concept that once you change someone's mind, it doesn't stick permanently.' In 2017, she conducted a study asking people whether they believed certain myths, like carrots improving your eyesight or liars moving their eyes in a certain direction.[72] The study found that they could often correct false beliefs, but the effect didn't necessarily last. 'If you get a correction, you might reduce your belief initially, but as time goes on you're going to re-believe in the initial misconception,' Swire-Thompson said. It seems repetition matters: new beliefs survived longer if people were reminded of the truth several times, rather than just given one correction.[73]

Thinking about the moral position of others. Having face-to-face interactions. Finding ways to encourage long-term change. All of these things can help improve persuasion. And it happens that they are also part of the deep canvassing approach advocated by the Los Angeles LGBT Center. Which brings us back to that dubious LaCour and Green paper. Although the study was retracted in 2015, the story didn't end there. The following year, David Brookman and Joshua Kalla – those two Berkeley researchers who'd found the problems in the original paper – published a new study.[74] This one focused on transgender rights. And this time they'd definitely collected the data.

Comparing deep canvassing with results from a control group, they'd found that a ten-minute conversation about transgender rights could noticeably reduce prejudice. It didn't matter whether the canvasser was transgender; the change in voters' opinion persisted regardless. The change in belief also seemed to be resistant to attacks. After a few weeks, the researchers showed people anti-transgender adverts from recent political campaigns. The ads initially swung opinions back against transgender people, but this reversion effect soon faded.

To ensure the research was completely transparent, Brookman and Kalla published all the data and code behind the analysis. It provided an optimistic epilogue to what had been an awkward few years for the research community. With the right approach, it was possible to change attitudes that many had believed were deeply ingrained. It showed that views don't necessarily spread in the way we assume they do, nor are people as fixed as we think they might be. When faced with apparent hostility, it seems there can be a lot to gain by trying something new.

4

Something in the air

'WE WERE IN A PLACE with real violence.' After a decade spent working on disease epidemics in Central and East Africa, Gary Slutkin had returned home to the United States. He'd chosen Chicago to be near his elderly parents and was struck by the extent of violent attacks in the city. 'It was surrounding, it was inescapable and so I just started to ask people what they were doing about it,' Slutkin said. 'And there wasn't anything that anybody was doing about this that made any sense to me.'[1]

It was 1994 and in the preceding year, there had been over eight hundred homicides in the city, including sixty-two children killed in gang violence. Even two decades later, homicide would still be the main cause of death for young adults in the state of Illinois.[2] Slutkin heard a range of explanations for the crisis, from nutrition and jobs to families and poverty. But the discussions often came back to a narrow set of solutions involving punishment. In his view, violence was what he called a 'stuck problem'. A physician by training, he'd seen similar situations in his work with infectious diseases like HIV/AIDS and cholera. Sometimes the thinking about a situation gets stuck for years. A strategy doesn't really work, but it doesn't change.

If violence were a stuck problem, it would need new thinking. 'You have to kind of start over,' Slutkin said. So he did what any public health researcher would do: he looked at maps and graphs, he asked questions, he tried to understand how violence was happening. And that's when he started noticing familiar patterns. 'The clustering seen in maps of killings in US cities resembles maps of cholera in Bangladesh,' he later wrote.[3] 'Historical graphs showing outbreaks of killing in Rwanda resembled graphs of cholera in Somalia.'

SUSANNAH ELEY LIKED TO GET her water delivered each day. After her husband had died, she'd moved from the bustle of London's Soho to leafy Hampstead. But she still preferred the water from the pump in town. She thought it tasted better.

One August day in 1854, Eley's niece visited her from the neighbouring borough of Islington. Within a week, they would both be dead. The culprit was cholera, an aggressive disease that causes diarrhea and vomiting. Left untreated, up to half of people with severe symptoms will die. The same day that Eley died from cholera, there were 127 other deaths from the disease, most of them in Soho. By the end of September, the outbreak would have claimed over six hundred lives in London. In this era before Koch's work on germ theory, the biology of cholera was still a mystery. 'We know nothing; we are at sea in a whirlpool of conjecture,' wrote Thomas Wakley, founder of *The Lancet* medical journal, the year before the outbreak started. People were starting to realise that diseases like smallpox and measles were contagious, somehow spreading from person to person, but cholera seemed to be something else. Most believed the 'miasma theory', which said that cholera spread through bad smells in the air.[4]

But not John Snow. Originally from Newcastle, Snow had investigated his first cholera outbreak in 1831 as an eighteen-year-old medical apprentice. Even then, he'd noticed some odd patterns. People who should have been at risk from bad air weren't getting ill, and people who supposedly weren't at risk were. Snow eventually moved to London, building up a reputation as a talented anaesthetist, with Queen Victoria among his patients. However, when a cholera outbreak hit the city in 1848, he revived his old investigations. Who was catching the disease? When were they getting ill? What linked the cases? The following year, Snow published an article with a new theory: the disease spread from one person to another through contaminated water. The realisation had finally come when he noticed that patients would often share the same water company. It was a remarkable insight, not least because Snow had no idea it was actually microscopic bacteria that were casting cholera's enormous shadow.

The 1854 Soho outbreak would prove a good match for Snow's theory. There were the workers at the local brewery, with their diet of ale and imported water, who didn't get sick. Then there was Susannah Eley and her niece, who had their water shipped from Soho to Hampstead and fell ill. As the outbreak grew, Snow decided it was time to intervene. Public health in Soho fell under the responsibility of a local Board of Guardians. He turned up uninvited at one of their meetings and presented his arguments. The board didn't fully believe his explanation, but decided to remove the pump handle all the same. The outbreak ended soon afterward.

Three months later, Snow wrote up his theory in more detail. The report included what would become his most famous illustration: a map of Soho, with black rectangles showing each of the cholera cases. The cases clustered around Broad Street, near the pump. It was a pioneering work of abstraction, removing unnecessary details and diversions. Whereas abstract artists like Malevich and Mondrian would later paint blocks of colour to shun reality, Snow's shapes brought cholera into focus.[5] His rectangles made a previously invisible truth – the source of infection – tangible.

Yet on its own, the map was not clear evidence that the water

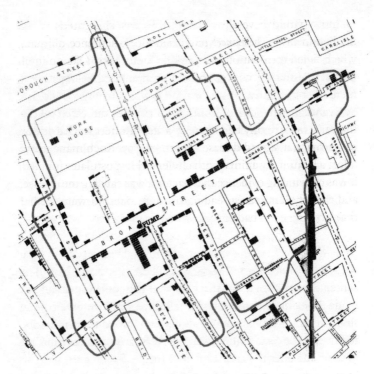

Snow's updated cholera map of Soho
Credit: John Snow Archive & Research Companion. The mark
on the right-hand side is a tear in the original page

was responsible. If the cholera outbreak had been the result of
bad air around Broad Street, the pattern would have looked
much the same. So Snow produced a second map, with a crucial
addition. As well as plotting the cases, he worked out how long
it would take to walk to different pumps, drawing a line to show
the places for which the Broad Street pump was nearest. It illus-
trated the areas that would be most at risk if the pump were to
blame. Just as his theory suggested, this was also where most
cases were appearing.

Snow would never live to see his ideas vindicated. When he died in 1858, *The Lancet* published a two-sentence obituary, which failed to mention his work on outbreaks. Like an intellectual miasma, the concept of bad air continued to linger in the medical community.

Eventually the idea of contagious cholera did catch on. By the early 1890s, many had come to accept Robert Koch's notion of germs that spread disease. Then, in 1895, Koch managed to infect a laboratory animal with cholera.[6] His postulates fulfilled, it was convincing evidence that bacteria was causing the disease, and that cholera was spreading through infected water rather than coming from bad air. Snow had been right.

WE NOW THINK ABOUT infectious diseases in terms of germs rather than miasma, but Gary Slutkin argues that we haven't made the same progress in our analysis of violence. 'We're very stuck in moralism – who's good, who's bad.' He points out that many societies are highly punitive; they haven't really shifted in their attitudes to violence for centuries. 'I really feel like I'm living in the past.'

Although biology has moved on from the idea of bad air, debate around crime still focuses on bad people. Slutkin thinks this is in part because contagious violence is less intuitive than disease. 'Here you don't actually have an invisible microorganism that you can at least show somebody under the microscope.' However, the parallels between infectious disease and violence seemed clear to him. 'I remember an epiphany when I asked someone "what's the greatest determinant of violence? What's the greatest predictor?" And the answer was "a preceding violent event".' In his mind, it was an obvious sign of contagion. Which made him wonder: perhaps methods used to control infectious diseases could be applied to violence too?

There are several similarities between outbreaks of disease

and violence. One is the lag between exposure and symptoms. Just like an infection, violence can have an incubation period; we might not see symptoms straight away. Sometimes a violent event will lead to another one soon after: for example, it might not take long for one gang to retaliate against another. On other occasions it may take much longer for knock-on effects to emerge. In the mid-1990s, epidemiologist Charlotte Watts worked with the World Health Organization (WHO) to set up a major study of domestic violence against women.[7] Watts had trained as a mathematician before moving into disease research, focusing on HIV. As her work on HIV developed, she started to notice that violence against women was influencing disease transmission because it affected their ability to have safe sex. But this revealed a much bigger problem: nobody really knew how common such violence was. 'Everybody agreed that we needed population data,' she said.[8]

The WHO study was the result of Watts and her colleagues applying public health ideas to the issue of domestic violence. 'A lot of previous research treated it as a police issue or focused on psychological drivers of violence,' she said. 'Public health people ask, "What's the big picture? What does the evidence say about individual, relationship and community risk factors?"' Some have suggested that domestic violence is completely context or culture specific, but this isn't necessarily the case. 'There are some really common elements that consistently come out,' Watts said, 'like exposure to violence in childhood.'

In most of the locations in the WHO study, at least one in four women had previously been physically abused by a partner. Watts has noted that violence can follow what's known in medicine as a 'dose-response effect'. For some diseases, the risk of illness can depend on the dose of pathogen a person is exposed to, with a small dose less likely to cause severe illness. There's evidence of a similar effect in relationships. If a man or woman has a history involving violence, it increases the chance of domestic violence in their future relationships. And if both members of the relationship have a history of violence, this risk increases even further. This isn't to say that people with a history involving violence will always have a violent future; like many infections, exposure to violence won't necessarily lead to symptoms later on. But like infectious diseases, there are a number of factors – in our backgrounds, in our lifestyles, in our social interactions – that can increase the risk of an outbreak.[9]

Another notable feature of disease outbreaks is that cases tend to cluster together in a certain location, with infections appearing over a short period of time. Think about that cholera outbreak in Broad Street, with cases clustered around the pump. We can find similar patterns when looking at violent acts. For centuries, people have reported localised clusters of self-harm and suicide: in schools, in prisons, in communities.[10] However, clustering of suicides doesn't necessarily mean contagion is happening.[11] As we saw with social contagion, people may behave in the same way for another reason, like some shared feature of their environment. One way to exclude this possibility is to look at the aftermath of high-profile deaths; a member of the public is more likely to hear about the suicide of a well-known person than the other way around. In 1974, David Phillips published a landmark paper examining media coverage of suicides. He found that when British and American newspapers ran a front-page story about a suicide, the number of such deaths in the local area tended to increase immediately afterwards.[12] Subsequent studies have found similar patterns with media reports, suggesting that suicide can be transmitted.[13] In response, WHO have published guidelines for responsible reporting of suicides. Media outlets should provide information about where to seek help, while avoiding sensational headlines, details about the method involved, and suggestions that the suicide was a solution to a problem.

Unfortunately, outlets often ignore these guidelines. Researchers at Columbia University noted a 10 per cent rise in suicides in the months following the death of comedian Robin Williams.[14] They pointed to a potential contagion effect, given that many media reports about Williams' death did not follow WHO guidelines, and the largest increase in suicides occurred in middle-aged men using the same method as Williams. There can be a similar effect with mass shootings; one study estimated that for every ten US mass shootings, there are two additional shootings as a result of social contagion.[15]

Because there is often an immediate rise in suicides and shootings following such media reports, it suggests that the delay between one contagious event and another – known in epidemiology as the 'generation time' – is relatively short. Some clusters of suicides have involved multiple deaths over a matter of weeks: in 1989 there was an outbreak of suicides at a Pennsylvania high school, which saw nine attempts in eighteen days. If these events were the result of contagion, the generation time may in some cases have been only a few days.[16]

Clustering is common with other types of violence too. In 2015, a quarter of US gun murders were concentrated in neighborhoods that made up less than 2 per cent of the country's overall population.[17] When Gary Slutkin and his colleagues set out to tackle violence as if it were an outbreak, it was neighbourhoods like these that they planned to target. They called the initial programme 'CeaseFire'; this would later evolve into a larger organisation called Cure Violence. In those early days, it took a while to work out precisely what approach they should use. 'We took five years of strategy development before we put a single thing on the street,' Slutkin said. The Cure Violence method would end up having three parts. First, the team hires 'violence interrupters' who can spot potential conflicts and intervene to stop the transmission of violence. Someone might end up in hospital with a gunshot wound, for example, and an interrupter will step in to talk their friends out of a retaliatory attack. Second, Cure Violence identifies who is at greatest risk of violence, using outreach workers to encourage a change in attitudes and behaviour. This can include help with things like job hunting or drug treatment. Finally, the team works to change social norms about guns in the wider community. The idea is to have a range of voices speaking out against a culture of violence.

Interrupters and outreach workers are recruited directly from the affected communities; some are former criminals or gang members. 'We hire workers who are credible with that population,' said Charlie Ransford, Cure Violence's Director of Science and Policy. 'To change people's behaviour and talk them out of doing something it helps if you have an understanding of where they're coming from, and they feel like you have an understanding and maybe even know you or know someone who knows you.'[18] This is another idea familiar in the world of infectious diseases: HIV programs will often recruit former sex workers to help change behaviour among workers who are still at high risk.[19]

The first Cure Violence project started in 2000, in West Garfield Park in Chicago. Why did they pick that location? 'It was the most violent police district in the country at the time,' Slutkin said. 'It has always been my bias – as it is for many epidemiologists – to head for the middle of the epidemic, because it's your best test and you can affect the greatest impact.' One year after the programme started, shootings in West Garfield Park had dropped by about two thirds. The change had been rapid, with interrupters breaking the chains of violence from one person to another. So what is it about these transmission chains that makes interruption possible?

LATE ON A SUNDAY AFTERNOON in May 2017, two gang members emerged from an alleyway in Chicago's Brighton Park neighbourhood. They were carrying assault rifles. The pair would end up shooting ten people, killing two of them. It was retaliation for a gang-related murder earlier in the day.[20]

Shootings in Chicago are often linked like this. Andrew Papachristos, a sociologist at Yale University, has spent several years studying patterns of gun violence in the city. A native of Chicago, he'd noticed that shootings were frequently tied to social contacts. Victims would often know each other, having previously been arrested together. Of course, just because two people are connected and share a characteristic – like involvement in a shooting – it doesn't necessarily mean that contagion is involved. It might be down to the environment they share, or because people tend to associate with those who have similar characteristics (i.e. homophily).[21]

To investigate further, Papachristos and his collaborators obtained data from the Chicago Police Department on everyone who'd been arrested between 2006 and 2014.[22] In total, there were over 462,000 people in the dataset. Using this information, they plotted a 'co-offending network' of people who'd previously been arrested at the same time. Many of the individuals hadn't ever been arrested with someone else, but there was a large group who could be linked together through a series of co-offending events. Overall this group included 138,000 people, or about a third of the dataset.

Papachristos's team started by checking whether homophily or environmental factors could explain the observed patterns of gun violence. They found that it was unlikely: many shootings occurred in a linked way that couldn't be explained by homophily or environment, suggesting contagion was responsible. Having identified the shootings that were likely due to contagion, the team carefully reconstructed the chains of transmission between one shooting and the next. They estimated that for every 100 people who were shot, contagion would result in 63 follow-up attacks. In other words, gun violence in Chicago had a reproduction number of about 0.63.

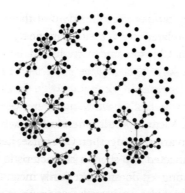

Fifty simulated outbreaks of shootings, based on the dynamics
of violence contagion in Chicago. Dots show shootings,
with (grey) arrows indicating follow-up attacks. Although
there are some superspreading events, most outbreaks
involve a single shooting and no onward transmission.

If the reproduction number is below one, it means that an
outbreak might spark but it rarely lasts very long. The Yale
team identified over four thousand outbreaks of gun violence in
Chicago, but most were small. The vast majority consisted of a
single shooting, with no additional contagion. However, occasionally the outbreaks were much larger; one included almost
five hundred linked shootings. When we see these highly variable
outbreak sizes, it suggests that transmission is driven by superspreading events. Analysing the outbreak data from Chicago in
more detail, I estimated that transmission of gun violence was
highly concentrated. It's likely that fewer than 10 per cent of
shootings led to 80 per cent of follow-up attacks.[23] Just like disease
transmission – which can be similarly influenced by superspreading – most shootings didn't lead to any additional contagion.

The chains of transmission in Chicago also revealed the speed
of transmission. On average, the generation time between one
shooting and another was 125 days. Despite the attention given

to dramatic retaliations like the Brighton Park attack in May 2017, it seems there are a lot of slower-burning feuds out there that have historically gone undetected.

These networks of shootings help explain why the Cure Violence approach is possible. Let's start with the fact that we can study the networks at all: if we want to control an outbreak, it helps if we can identify potential routes of transmission. Slutkin has compared violence interruption to the methods used to control smallpox outbreaks. As smallpox was nearing eradication in the 1970s, epidemiologists used 'ring vaccination' to stamp out the final few sparks of infection. When a new disease case appeared, teams would track down people the infected may have come into contact with, such as family members and neighbours, as well as these people's contacts. They would then vaccinate people within this 'ring', preventing the smallpox virus spreading any further.[24]

Smallpox had three features that worked in health teams' favour. To spread from one person to another the disease generally required fairly long face-to-face interactions. This meant teams could identify who was most at risk. In addition, the generation time for smallpox was a couple of weeks; when a new case was reported, teams had enough time to go and vaccinate before more cases appeared. Finally, cases developed a distinctive rash, making them easy to spot. The spread of gun violence shares these features: shootings are noticeable events, violence is often transmitted through known social links, and the gap between one shooting and the next is long enough for interrupters to intervene. If shootings went undetected, were more random, or the gap between them was always much shorter, violence interruption wouldn't be so effective. (By comparison, COVID-19 is difficult to control because it lacks some of these features: people can spread infection without clear symptoms, and the generation time is relatively short at around five days).[25]

An independent evaluation of Cure Violence by the US

National Institute for Justice found a substantial drop in shootings in areas where the programme had been introduced. It can be tough to assess the precise impact of anti-violence programmes, because violence may have already been declining for some other reason. But violence hadn't declined as much in comparable areas of Chicago, suggesting that Cure Violence was in fact behind the reduction in shootings in many locations. In 2007, Cure Violence started working in Baltimore. When researchers at Johns Hopkins University later assessed the results, they estimated that in its first two years, the programme had prevented around thirty-five shootings and five homicides. Other studies have found similar reductions after the introduction of Cure Violence methods.[26]

Even so, the Cure Violence approach has not been free from criticism. Much of the scepticism has come from those in charge of existing approaches; in the past, there have been complaints from Chicago police about a lack of co-operation from interrupters. There have also been instances of violence interrupters being charged with other crimes. Such challenges are perhaps inevitable, given that the programme relies on having interrupters that are part of the communities at risk, rather than another branch of the police.[27] Then there's the timescale of social change. While stopping retaliatory attacks can have an immediate effect on violence, tackling the underlying social issues may take years.[28] The same is true with infectious diseases: we might be able to stop outbreaks, but we also need to think about underlying weaknesses in health systems that enabled them in the first place.

Building on their early work in Chicago, Cure Violence has expanded to other US cities, including Los Angeles and New York, as well as launching projects in countries like Iraq and Honduras. Public health approaches would also inspire a 'Violence Reduction Unit' in Glasgow, Scotland. Back in 2005, the city was named the murder capital of Europe. There were dozens of knife attacks a week, including numerous incidents of notorious 'Glasgow smiles' being slashed into people's cheeks. What's more, the violence was far more widespread than police figures suggested. When Karyn McCluskey, head of intelligence analysis at Strathclyde Police, looked at hospital records, it became clear that most incidents weren't even being reported.[29]

McCluskey's findings – and accompanying recommendations – led to the creation of the Violence Reduction Unit, which she would head up for the following decade. Borrowing techniques from Cure Violence and other US projects, such as Boston's Operation Ceasefire, the unit introduced a range of public health ideas to tackle the spread of violence.[30] This included interruption approaches, like monitoring A&E departments for victims of violence to discourage potential revenge attacks. It also involved helping gang members move into training and employment, while taking a tough stance against those who chose to continue with violence. There were longer-term measures too, like providing support for vulnerable children to halt the transmission of violence from generation to the next. Although there is still more to be done, the initial results have been promising; following its introduction, the unit has been linked with a major drop in violent crime.[31]

Since 2018, London has been working on a similar initiative to tackle what has been described as an 'epidemic' of knife crime in the city. If it is to succeed like Glasgow, it will require strong links between police, communities, teachers, health services, social workers, and the media. It will also need continued investment, given the often complex, deep-rooted nature of the problem. 'It's about putting money where your mouth is in terms of prevention, and understanding that you may not see a really quick return on it,' McCluskey told The Independent shortly before the London project launched.[32]

Sustaining investment can be tough for public health approaches. Despite growing acceptance elsewhere, funding for the original Cure Violence programme in Chicago has remained sporadic, with several cutbacks over the years. Slutkin said attitudes to violence are changing in many places, but not as easily as he would hope. 'It's frustratingly slow,' he said.

ONE OF THE BIGGEST CHALLENGES in public health is convincing people. It's not just a matter of showing a new approach works better than existing methods. It's also about advocating for that approach, presenting a compelling argument that can help turn statistical evidence into action.

In the world of public health advocacy, few have been as effective – or as pioneering – as Florence Nightingale. While John Snow was analysing cholera in Soho, Nightingale was surveying the illnesses faced by British troops fighting in the Crimean war. Nightingale had arrived in late 1854 to lead a team of nurses in the military hospitals. She found that soldiers were dying at an astonishing rate. It wasn't just the fighting that was killing them; it was infections like cholera, typhoid, typhus and dysentery. In fact, infections were the main source of death. During 1854, eight times more soldiers died from diseases than from battle wounds.[33]

Nightingale was convinced poor hygiene was to blame. Each night, she walked over six kilometres along the corridors of the wards, lamp in hand. Patients lay on filthy mattresses, rats hiding beneath, surrounded by walls covered in dirt. 'The clothes of those men were swarming with lice,' Nightingale noted, 'as thick as the letters on a page of print.' With her nurses, she set about cleaning up the wards. They made sure linens were laundered, bodies bathed, and walls washed. In March 1855, the British government sent a group of commissioners to the Crimea to tackle conditions in the hospitals. Whereas Nightingale had focused on hygiene, the commission worked on the buildings, improving ventilation and sewage systems.

Nightingale's work earned her fame back at home. Shortly after returning to England in summer 1856, Queen Victoria invited her to come to Balmoral to discuss her experiences in the Crimea. Nightingale used the meeting to push for a Royal Commission to examine the high death rates. What had really happened out there?

As well as contributing to the commission, Nightingale continued with her own research into the hospital data. This work accelerated after she met statistician William Farr at a dinner party that autumn. The two had very different backgrounds: Nightingale came from the upper class, with a name reflecting her childhood in Tuscany, while Farr had been raised in poverty in rural Shropshire, eventually studying medicine before moving into medical statistics.[34]

When it came to population data in the 1850s, Farr was the man to speak to. Alongside his work on outbreaks like smallpox, he had set up the first national system to collate data on things like births and deaths. However, he'd noticed that these raw statistics could be misleading. The total number of deaths in a particular area would depend on how many inhabitants there were, as well as factors like age: a town with an elderly population would generally have more deaths each year than a town full of young people. To solve this problem, Farr came up with a new measurement. Rather than study total deaths, he looked at the rate of death per thousand people, accounting for things like age. It meant he could compare different populations in a fair way. 'The death-rate is a fact; anything beyond this is an inference,' as Farr put it.[35]

Working with Farr, Nightingale applied these new methods to data from the Crimea. She showed that death rates in army hospitals were much higher than wards in Britain. She also measured the decline in disease after the health commissioners arrived in 1855. As well as producing tables of data, she took full advantage of a new trend in Victorian science: data visualisation.

Economists, geographers and engineers had increasingly used graphs and figures to make their work more accessible. Nightingale adapted these techniques, converting her key results into bar graphs and pie chart-like figures. Like Snow's maps, the graphics focused on the most important patterns, free of distractions. The visuals were clear and memorable, helping her message to spread.

In 1858, she published her analysis of health in the British Army as an 860-page book. Copies were shipped to leaders ranging from Queen Victoria and the Prime Minister to newspaper editors and European heads of state. Whether looking at hospitals or communities, Nightingale believed that nature followed predictable laws when it came to disease. She said those disastrous early months in Crimea happened because people ignored these laws. 'Nature is the same everywhere, and never permits her laws to be disregarded with impunity.' She was also adamant about what had caused the problems. 'The three things which all but destroyed the army in Crimea were ignorance, incapacity, and useless rules.'[36]

Nightingale's advocacy sometimes made Farr nervous. He warned her against focusing too heavily on messages rather than data. 'We do not want impressions,' he said. 'We want facts.'[37] Whereas Nightingale wanted to suggest explanations for the cause of the deaths, Farr believed the job of a statistician was simply to report what had happened, rather than speculating about why. 'You complain that your report would be dry,' he once told her. 'The drier the better. Statistics should be the driest of all reading.'

Nightingale used her writing to campaign for change, but she'd never wanted to be just a writer. When she first decided to train as a nurse in the 1840s, it came as a surprise to her wealthy, well-connected family, who'd expected her to pursue the more traditional role of wife and mother. A friend suggested that she could still pursue a literary career alongside this role.

Nightingale was not interested. 'You ask me why I do not write something,' she replied. 'I think one's feelings waste themselves in words; they ought all to be distilled into actions and into actions which bring results.'[38]

When it comes to improving health, actions need to be grounded in good evidence. Today, we routinely use data analysis to show how much health varies, why that might be, and what needs to be done about it. Much of this evidence-based approach can be traced to statisticians like Farr and Nightingale. As she saw it, people generally had little grasp of what controlled infections and what didn't. In some cases, hospitals may well have increased people's risk of disease. 'These institutions, created for the relief of human distress, positively do not know whether they relieve it or not,' as she put it.[39]

Nightingale's research was highly respected by her scientific contemporaries, including statistician Karl Pearson. In the public mind, she was the 'lady with the lamp', a nurse who cared for soldiers and in turn made people sympathetic to her cause. But Pearson argued that mere sympathy doesn't lead to change; it requires knowledge of management and administration, as well as an ability to interpret information. He said this was where Nightgale excelled. 'Florence Nightingale believed – and in all the actions of her life acted upon that belief – that the administrator could only be successful if he were guided by statistical knowledge.'[40]

ACCORDING TO CARL BELL, a public health specialist at the University of Chicago, three things are required to stop an epidemic: an evidence base, a method for implementation, and political will.[41] Yet when it comes to gun violence, the US has struggled even with the first step. The US Centers for Disease Control and Prevention (CDC), who would usually take the lead on public health matters, have done very little research into the problem in the past two decades.

Without a doubt, the US is a big outlier when it comes to guns. In 2010, young American adults were almost fifty times more likely to die in a shooting than their peers in other high-income countries. The media tend to focus on mass shootings, which often involve assault weapons, but the problem of gun deaths is far more widespread than this. In 2016, mass shootings – defined as four or more people being shot – made up just 3 per cent of US gun homicides.[42]

So why hasn't the CDC done more research into gun violence? The main reason is the 1996 Dickey Amendment, which stipulates that 'none of the funds made available for injury prevention and control at the CDC may be used to advocate or promote gun control.' Named after Republican congressman Jay Dickey, the amendment followed a series of disagreements about gun research in the US. In the run up to the vote, Dickey and his colleagues had clashed with Mark Rosenberg, director of the National Center for Injury Prevention and Control at the CDC. They claimed that Rosenberg, who co-chaired a firearms working group, was trying to present guns as a 'public health menace' (the phrase actually came from a *Rolling Stone* journalist who'd interviewed Rosenberg about gun violence).[43]

Rosenberg had contrasted gun research to the progress made in reducing car-related deaths, an analogy later used by Barack Obama during his presidency. 'With more research, we could further improve gun safety just as with more research we've reduced traffic fatalities enormously over the last 30 years,' Obama said in 2016. 'We do research when cars, food, medicine, even toys harm people so that we make them safer. And you know what, research, science, those are good things. They work.'[44]

Cars have become much safer, but the industry was initially reluctant to accept suggestions that their vehicles needed improvements. When Ralph Nader published his 1965 book *Unsafe at Any Speed*, which presented evidence of dangerous design flaws, car companies attempted to smear him. They got private detectives to track his movements and hired a prostitute to try and seduce him.[45] Even the book's publisher, Richard Grossman, was sceptical about the message. He thought it would be hard to market and probably wouldn't sell very well. 'Even if every word in it is true and everything about it is as outrageous as he says,' Grossman later recalled, 'do people want to read about that?'[46]

It turned out that they did. *Unsafe at Any Speed* became a best-seller and calls to improve road safety grew, leading to seat belts and eventually features like airbags and antilock brakes. Even so, it had taken a while for the evidence to accumulate prior to Nader's book. In the 1930s, many experts thought it was safer to be thrown from a car during an accident, rather than be stuck inside.[47] For decades, manufacturers and politicians weren't that interested in car safety research. After the publication of *Unsafe at Any Speed*, that changed. In 1965, a million miles of car travel came with a 5 per cent chance of death; by 2014 this had dropped to 1 per cent.

Before he died in 2017, Jay Dickey indicated that his views on gun research had shifted. He believed the CDC needed to look at gun violence. 'We need to turn this over to science and take it away from politics,' he told the *Washington Post* in 2015.[48] In the years following their 1996 clash, Dickey and Mark Rosenberg had become friends, taking time to listen and find common ground on the need for gun research. 'We won't know the cause of gun violence until we look for it,' they would later write in a joint opinion piece.

Despite constraints on funding, some evidence about gun violence is available. In the early 1990s, before the Dickey Amendment, CDC-funded studies found that having a gun in the home increased the risk of homicide and suicide. The latter finding was particularly notable, given that around two-thirds of gun deaths in the US are from suicide. Opponents of this research have argued that such suicides might have occurred anyway, even if guns hadn't been present.[49] But easy access to deadly methods can make a difference for what are often impulse decisions. In 1998, the UK switched from selling paracetamol in bottles to blister packs containing up to thirty-two tablets. The extra effort involved with blister packs seemed to deter people; in the decade after the packs were introduced, there was about a 40 per cent reduction in deaths from paracetamol overdoses.[50]

Unless we understand where the risk lies, it's very difficult to do anything about it. This is why research into violence is needed. Seemingly obvious interventions may turn out to have little effect in reality. Likewise, there may be policies – like Cure Violence – that challenge existing approaches, but have the potential to reduce gun-related deaths. 'Like motor vehicle injuries, violence exists in a cause-and-effect world; things happen for predictable reasons,' wrote Dickey and Rosenberg in 2012.[51] 'By studying the causes of a tragic – but not senseless – event, we can help prevent another.'

It's not just gun violence that we need to understand. So far, we've looked at frequently occurring events like shootings and domestic violence, which means there is – in theory, at least – a lot of data to study. But sometimes crime and violence happen as a one-off event, spreading rapidly through a population with devastating consequences.

ON THE EVENING OF SATURDAY 6 August 2011, London descended into what would become the first of five nights of looting, arson and violence. Two days earlier, police had shot and killed a suspected gang member in Tottenham, North London, sparking protests that evolved into riots and spread across the city. There would also be rioting in other UK cities, from Birmingham to Manchester.

Crime researcher Toby Davies was living in the London

district of Brixton at the time.[52] Although Brixton avoided the violence on the first night of the riots, it would end up being one of the worst affected areas. In the months following the riots, Davies and his colleagues at University College London decided to pick apart how such disorder could develop.[53] Rather than trying to explain how or why a riot starts, the team instead focused on what happens once it gets underway. In their analysis, they divided rioting into three basic decisions. The first was whether a person would participate in the riot or not. The researchers assumed this depended on what was happening nearby – much like a disease epidemic – as well as local socioeconomic factors. Once someone decided to participate, the second decision involved where to riot. Because a lot of the rioting and looting was concentrated in retail areas, the researchers adapted an existing model for how shoppers flow into such locations (several media outlets described the London riots as 'violent shopping'[54]). Finally, their model included the possibility of arrest once a person arrived at the rioting site. This depended on the relative number of rioters and police, a metric Davies referred to as 'outnumberedness'.

The model could reproduce some of the broad patterns seen during the 2011 riots – such as the focus on Brixton – but it also showed the complexity of these types of events. Davies points out that the model was only a first step; there's a lot more that needs to be done in this area of research. One big challenge is the availability of data. In their analysis, the UCL team only had information on the number of arrests for riot-related offences. 'As you can imagine, it's a very small and very biased subsample,' Davies said. 'It doesn't capture who could potentially engage in rioting.' In 2011, the rioters were also more diverse than might be expected, with groups transcending long-standing local rivalries. Still, one of the benefits of a model is that it can explore unusual situations and potential responses. For frequent crimes like burglary, police can introduce control measures, see what happens, then refine their strategy. However, this approach isn't possible for rare events, which might only spark now and again. 'Police don't have riots to practise on every day,' Davies said.

For a riot to start, there need to be at least some people willing to join. 'You cannot riot on your own,' as crime researcher John Pitts put it. 'A one-man riot is a tantrum.'[55] So how does a riot grow from a single person? In 1978, Mark Granovetter published a now classic study looking at how trouble might take off. He suggested that people might have different thresholds for rioting: a radical person might riot regardless of what others were doing, whereas a conservative individual might only riot if many others were. As an example, Granovetter suggested we imagine 100 people hanging around in a square. One person has a threshold of 0, meaning they'll riot (or tantrum) even if nobody else does; the next person has a threshold of 1, so they will only riot if at least one other person does; the next person has a threshold of 2, and so on, increasing by one each time. Granovetter pointed out that this situation would lead to an inevitable domino effect: the person with a 0 threshold would start rioting, triggering the person with a threshold of 1, which would trigger the person with a threshold of 2. This would continue until the entire crowd was rioting.

But what if the situation were slightly different? Say the person with a threshold of 1 had a threshold of 2. This time, the first person would start rioting, but there would be nobody else with a low enough threshold to be triggered. Although the crowds in each situation are near identical, the behaviour of one person could be the difference between a riot and a tantrum. Granovetter suggested personal thresholds could apply to other forms of collective behaviour too, from going on strike to leaving a social event.[56]

The emergence of collective behaviour can also be relevant to counter-terrorism. Are potential terrorists recruited into an existing hierarchy, or do they form groups organically? In 2016, physicist Neil Johnson led an analysis looking at how support for the so-called Islamic State grew online. Combing through discussions on social networks, his team found that supporters aggregated in progressively larger groups, before breaking apart into smaller ones when the authorities shut them down. Johnson has compared the process to a school of fish splitting and reforming around predators. Despite gathering into distinct groups, Islamic State supporters didn't seem to have a consistent hierarchy.[57] In their studies of global insurgency, Johnson and his collaborators have argued that these collective dynamics in terrorist groups could explain why large attacks are so much less frequent than smaller ones.[58]

Although Johnson's study of Islamic State activity aimed to understand the ecosystem of extremism – how groups form, grow, and dissipate – the media preferred to focus on whether it could accurately predict attacks. Unfortunately, predictions are probably still beyond the reach of such methods. But at least it was possible to see what the underlying methods were. According to J.M. Berger, a fellow at George Washington University who researches extremism, it's rare to see such transparent analysis of terrorism. 'There are a lot of companies that claim to be able to do what this study is claiming,' he told the *New York Times* after the study was published, 'and a lot of those companies seem to me to be selling snake oil.'[59]

PREDICTION IS A DIFFICULT BUSINESS. It's not just a matter of anticipating the timing of a terrorist attack; governments also have to consider the method that may be used, and the potential impact that method will have. In the weeks following the 9/11 attacks in 2001, several people in the US media and Congress received letters containing toxic anthrax bacteria. It led to five deaths, raising concerns that other bioterrorist attacks may follow.[60] One of the top threats was thought to be smallpox. Despite having been eradicated in the wild, samples of the virus were still stored in two government labs, one in the US and one in Russia. What if other, unreported, smallpox viruses were out there and fell into the wrong hands?

Using mathematical models, several research groups tried to estimate what might happen if terrorists released the virus into a human population. Most concluded that an outbreak would grow quickly unless pre-emptive control measures were in place. Soon after, the US Government decided to offer half a million healthcare workers vaccination against the virus. There was limited enthusiasm for the plan: by the end of 2003, fewer than 40,000 workers had opted for the vaccine.

In 2006, Ben Cooper, then a mathematical modeller at the UK Health Protection Agency, wrote a high-profile paper critiquing the approaches used to assess the smallpox risk. He titled it 'Poxy Models and Rash Decisions'. According to Cooper, several models included questionable assumptions, with one particularly prominent example. 'Collective eyebrows were raised when the Centers for Disease Control's model completely neglected contact tracing and forecast 77 trillion cases if the epidemic went unchecked,' he noted. Yes, you read that correctly. Despite there being fewer than 7 billion people in the world at the time, the model had assumed that there were an infinite number of susceptible people that could become infected, which meant transmission would continue indefinitely. Although the CDC researchers acknowledged it was a major simplification, it was bizarre to see an outbreak study make an assumption that was so dramatically detached from reality.[61]

Still, one of the advantages of a simple model is that it's usually easy to spot when – and why – it's wrong. It's also easier to debate the usefulness of that model. Even if someone has limited experience with mathematics, they can see how the assumptions influence the results. You don't need to know any calculus to notice that if researchers assume a high level of smallpox transmission and an unlimited number of susceptible people, it can lead to an unrealistically large epidemic.

As models become more complicated, with lots of different features and assumptions, it gets harder to identify their flaws. This creates a problem, because even the most sophisticated mathematical models are a simplification of a messy, complex reality. It's analogous to building a child's model train set. No matter how many features are added – miniature signals, numbers on the carriages, timetables full of delays – it is still just a model. We can use it to understand aspects of the real thing, but there will always be some ways in which the model will differ from the true situation. What's more, additional features may not make a model better at representing what we need it to. When it comes to building models, there is always a risk of confusing detail with accuracy. Suppose that in our train set all the trains are driven by intricately carved and painted zoo animals. It might be a very detailed model, but it's not a realistic one.[62]

In his critique, Cooper noted that other, more detailed smallpox models had come to similarly pessimistic conclusions about the potential for a large outbreak. Despite the additional detail, though, the models still contained an unrealistic feature: they had assumed that most transmission occurred before people developed the distinctive smallpox rash. Real life data suggested otherwise, with the majority of transmission happening after the rash appeared. This would make it much easier to spot who was infectious, and hence control the disease through quarantine rather than requiring widespread vaccination.

From disease epidemics to terrorism and crime, forecasts can help agencies plan and allocate resources. They can also help draw attention to a problem, persuading people that there is a need to allocate resources in the first place. A prominent example of such analysis was published in September 2014. In the midst of the Ebola epidemic that was sweeping across several parts of

West Africa, the CDC announced that there could be 1.4 million cases by the following January if nothing changed.[63] Viewed in terms of Nightingale-style advocacy, the message was highly effective: the analysis caught the world's attention, attracting widespread media coverage. Like several other studies around that time, it suggested that a rapid response was needed to control the epidemic in West Africa. But the CDC estimate soon attracted criticism from the wider disease research community.

One issue was the analysis itself. The CDC group behind the number was the same one that had come up with those small-pox estimates. They'd used a similar model, with an unlimited number of susceptible people. If their Ebola model had run until April 2015, rather than January, it would have estimated over 30 million future cases, far more than the combined populations of the countries affected.[64] Many researchers questioned the appropriateness of using a very simple model to estimate how Ebola might be spreading five months later. I was one of them. 'Models can provide useful information about how Ebola might spread in the next month or so,' I told journalists at the time, 'but it is near impossible to make accurate longer-term forecasts'.[65]

To be clear, there are some very good researchers within the wider CDC, and the Ebola model was just one output from a large research community there. But it does illustrate the challenges of producing and communicating high profile outbreak analysis. One problem with flawed predictions is that they reinforce the idea that models aren't particularly useful. If models produce incorrect forecasts, the argument goes, why should people pay attention to them?

We face a paradox when it comes to forecasting outbreaks. Although pessimistic weather forecasts won't affect the size of a storm, outbreak predictions can influence the final number of cases. If a model suggests the outbreak is a genuine threat, it may trigger a major response from health agencies. And if this brings the outbreak under control, it means the original forecast will be wrong. It's therefore easy to confuse a useless forecast (i.e. one that would never have happened) with a useful one, which would have happened had agencies not intervened. Similar situations can occur in other fields. In the run up to the year 2000, governments and companies spent hundreds of billions of dollars globally to counter the 'Millennium bug'. Originally a feature to save storage in early computers by abbreviating dates, the bug had propagated through modern systems. Because of the efforts to fix the problem, the damage was limited in reality, which led many media outlets to complain that the risk had been overhyped.[66]

Strictly speaking, the CDC Ebola estimate avoided this problem because it wasn't actually a forecast; it was one of several scenarios. Whereas a forecast describes what we think will happen in the future, a scenario shows what could happen under a specific set of assumptions. The estimate of 1.4 million cases assumed the epidemic would continue to grow at the exact same rate. If disease control measures were included in the model, it predicted far fewer cases. But once numbers are picked up, they can stick in the memory, fueling scepticism about the kinds of models that created them. 'Remember the 1 million Ebola cases predicted by CDC in fall 2014,' tweeted Joanne Liu, International President of Médecins Sans Frontières (MSF), in response to a 2018 article about forecasting.[67] 'Modeling has also limits.'

Even if the 1.4 million estimate was just a scenario, it still implied a baseline: if nothing had changed, that is what would have happened. During the 2013–2016 epidemic, almost 30,000 cases of Ebola were reported across Liberia, Sierra Leone and Guinea. Did the introduction of control measures by Western health agencies really prevent over 1.3 million cases?[68]

In the field of public health, people often refer to disease control measures as 'removing the pumphandle.' It's a nod to

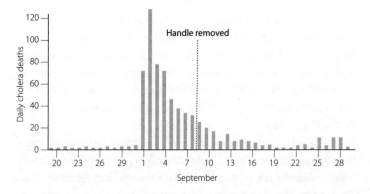

Soho cholera outbreak, 1854

John Snow's work on cholera, and the removal of the handle on the Broad Street pump. There's just one problem with this phrase: when the pumphandle came off on 8 September 1854, London's cholera outbreak was already well in decline. Most of the people at risk had either caught the infection already, or fled the area. If we're being accurate, 'removing the pumphandle' should really refer to a control measure that's useful in theory, but delivered too late.

By the time some of the largest Ebola treatment centres opened in late 2014, the outbreak was already slowing down, if not declining altogether.[69] Yet in some areas, control measures did coincide with a fall in cases. It's therefore tricky to untangle the exact impact of these measures. Response teams often introduced several measures at once, from tracing infected contacts and encouraging changes in behaviour to opening treatment centres and conducting safe burials. What effect did international efforts actually have?

Using a mathematical model of Ebola transmission, our group estimated that the introduction of additional treatment beds – which isolated cases from the community and thereby reduced transmission – prevented around 60,000 Ebola cases in Sierra Leone between September 2014 and February 2015. In some districts, we found that the expansion of treatment centres could explain the entire outbreak decline; in other areas, there was evidence of an additional reduction in transmission in the community. This could have reflected other local and international control efforts, or perhaps changes in behaviour that were occurring anyway.[70]

Historical Ebola outbreaks have shown how important behaviour changes can be for outbreak control. When the first reported outbreak of Ebola started in the village of Yambuku, Zaire (now the Democratic Republic of the Congo) in 1976, the infection sparked in a small local hospital before spreading to the community. Based on archive data from the original outbreak investigation, my colleagues and I estimated that the transmission rate in the community declined sharply a few weeks into the outbreak.[71] Much of the decline came before the hospital closed and before the international teams arrived. 'The communities where the outbreak continued to spread developed their own form of social distancing,' recalled epidemiologist David Heymann, who was part of the investigation.[72] Without doubt, the international response to Ebola in late 2014 and early 2015 helped prevent cases in West Africa. But at the same time, foreign organisations should be cautious about claiming too much credit for the decline of such outbreaks.

DESPITE THE CHALLENGES INVOLVED in producing forecasts, there is a large demand for them. Whether we're looking at the spread of infectious diseases or crime, governments and other organisations need evidence to base their future policies on. So how can we improve outbreak forecasts?

Generally, we can trace problems with a forecast back to either the model itself or the data that goes into it. A good rule of thumb is that a mathematical model should be designed around the data available. If we don't have data about the different transmission routes, for example, we should instead try to make simple but plausible assumptions about the overall spread. As well as making models easier to interpret, this approach also makes it easier to communicate what is unknown. Rather than grappling with a complex model full of hidden assumptions, people will be able to concentrate on the main processes, even if they're not so familiar with modelling.

Outside my field, I've found that people generally respond to mathematical analysis in one of two ways. The first is with suspicion. This is understandable: if something is opaque and unfamiliar, our instinct can be to not trust it. As a result, the analysis will probably be ignored. The second kind of response is at the other extreme. Rather than ignore results, people may have too much faith in them. Opaque and difficult is seen as a good thing. I've often heard people suggest that a piece of maths is brilliant because nobody can understand it. In their view, complicated means clever. According to statistician George Box, it's not just observers who can be seduced by mathematical analysis. 'Statisticians, like artists, have the bad habit of falling in love with their models,' he supposedly once said.[73]

We also need to think about the data we put into our analysis. Unlike scientific experiments, outbreaks are rarely designed: data can be messy and missing. In retrospect, we may be able to plot neat graphs with cases rising and falling, but in the middle of an outbreak we rarely have this sort of information. In December 2017, for example, our team worked with MSF to analyse an outbreak of diphtheria in refugee camps in Cox's Bazar, Bangladesh. We received a new dataset each day. Because it took time for new cases to be reported, there were fewer recent cases in each of these datasets: if someone fell ill on a Monday, they generally wouldn't show up in the data until Wednesday or Thursday. The epidemic was still going, but these delays made it look like it was almost over.[74]

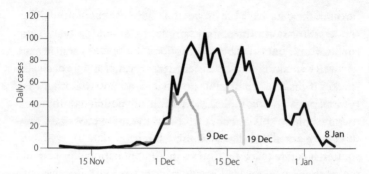

Diphtheria outbreak in Cox's Bazar Bangladesh, 2017–18. Each line
shows the number of new cases on a given day, as reported in the
database as it appeared on 9 December, 19 December and 8 January.

Data: Finger et al., 2019

Although outbreak data can be unreliable, it doesn't mean it's
unusable. Imperfect data isn't necessarily a problem if we know
how it's imperfect, and can adjust accordingly. For example,
suppose your watch is an hour slow. If you aren't aware of this,
it will probably cause you problems. But if you know about the
delay, you can make a mental adjustment and still be on time.
Likewise, if we know the delay in reporting during an outbreak,
we can adjust how we interpret the outbreak curve. Such 'now-
casting', which aims to understand the situation as it currently
stands, is often necessary before forecasts can be made.

Our ability to nowcast will depend on the length of the delay
and the quality of data available. Many infectious disease outbreaks last weeks or months, but other outbreaks can occur on
much longer timescales. Take the so-called opioid epidemic in
the US, in which a rising number of people are addicted to prescription painkillers, as well as illegal drugs like heroin. Drug
overdoses are now the leading cause of death for Americans
under the age of 55. As a result of these additional deaths,
average life expectancy in the US declined three years running

between 2015 and 2018. The last time that happened was the Second World War. Despite some aspects of the crisis being specific to the US, it isn't the only area at risk; opioid use has also been on the rise in places like the UK, Australia and Canada.[75]

Unfortunately, it's hard to track drug overdoses because it takes especially long to certify deaths as drug-related. Preliminary estimates for US overdose deaths in 2018 weren't released until July 2019.[76] Although some local-level data is available sooner, it can take a long time to build up a national picture of the crisis. 'We're always looking backwards,' said Rosalie Liccardo Pacula, a senior economist at the RAND Corporation, which specialises in public policy research. 'We aren't very good at being able to see what's happening immediately.'[77]

The US opioid crisis has received substantial attention in the twenty-first century, but Hawre Jalal and colleagues at the University of Pittsburgh suggest that the problem goes back much further. When they looked at data between 1979 and 2016, they found that the number of overdose deaths in the US grew exponentially during this period, with the death rate doubling every ten years.[78] Even when they looked at the state rather than national level, they found the same growth pattern in many areas. The consistency of the growth pattern was surprising given how much drug use has changed over the decades. 'This historical pattern of predictable growth for at least 38 years suggests that the current opioid epidemic may be a more recent manifestation of an ongoing longer-term process,' the researchers noted. 'This process may continue along this path for several more years into the future.'[79]

Yet drug overdose deaths only show part of the picture. They don't tell us about the events that led up to this point; a person's initial misuse of drugs may have started years earlier. This time lag happens in most types of outbreak. When people come into contact with an infection, there is usually a delay between being exposed and observing the effects of that exposure. For

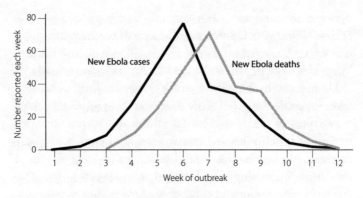

1976 Ebola outbreak in Yambuku
Data: Camacho et al., 2014

example, during that 1976 Ebola outbreak in Yambuku, people who were exposed to the virus often took a few days to become ill. For infections that were fatal, there was then another week or so between the illness appearing and death. Depending on whether we look at illnesses or deaths, we get two slightly different impressions of the outbreak. If we focus on newly ill Ebola cases, we'd say that the Yambuku outbreak peaked after six weeks; based on deaths, we'd put the peak a week later.

Both datasets are useful, but they're not measuring quite the same thing. The tally of new Ebola cases tells us what is happening to susceptible people – specifically, how many are getting infected – whereas the number of deaths shows what is happening to people who already have the infection. After the first peak, the two curves go in opposite directions for a week or so: cases fall while deaths are still rising.

According to Pacula, drug epidemics can be divided into similar stages. In the early stage of an outbreak, the number of users increases, as new people are exposed to drugs. In the case of opioids, exposure often starts with a prescription. It might be tempting to simply blame patients for taking too much medication, or doctors for overprescribing. But we must also consider the pharmaceutical companies who market strong opioids directly to doctors. And insurance companies, who are often more likely to fund painkillers than alternatives like physiotherapy. Our modern lifestyles also play a role, with rising chronic pain associated with increases in obesity and office-based work.

One of the best ways to slow an epidemic in its early stages is to reduce the number of people who are susceptible. For drugs, this means improving education and awareness. 'Education has been very important and very effective,' said Pacula. Strategies that reduce the supply of drugs can also help early on. Given the multitude of drugs involved in the opioid epidemic, this means targeting all potential routes of exposure, rather than one specific medication.

Once the number of new users peaks, we enter the middle stage of a drug epidemic. At this point, there are still a lot of existing users, who may be progressing towards heavier drug use, and potentially moving on to illegal drugs as they lose their access to prescriptions. Providing treatment and preventing heavy use can be particularly effective at this stage. The aim here is to reduce the overall number of users, rather than just preventing new addictions.

In the final stage of a drug epidemic, the number of new and existing users is declining, but a group of heavy users remains. These are the people who are most at risk, having potentially switched from prescription opioids to cheaper drugs like heroin.[80] But it's not as simple as cracking down on the illegal drug market in these later stages. The underlying problem of addiction is much deeper and wider than this. As Police Chief Paul Cell put it, 'America can't arrest its way out of the opioid epidemic'.[81] Nor is it just a matter of taking away access to prescription drugs. 'There's an addiction problem, and not just an opioid problem,' Pacula said. 'If you don't provide treatment when you're taking away the drug, you're basically encouraging them to go to anything else.' She pointed out that drug epidemics also come with a series of knock-on effects. 'Even if we get the issue of misuse of opioids under control, we have some very concerning long term trends that we haven't even started dealing with.' One is the effect on drug users' health. As people move from taking pills to injecting drugs, they face the risk of infections like hepatitis C and HIV. Then there is the wider social impact – on families, communities, and jobs – of having large numbers of people with drug addiction.

Because the success of different control strategies can vary between the three stages of a drug epidemic, it's crucial to know what stage we're currently in. In theory, it should be possible to work this out by estimating the annual numbers of new users, existing users, and heavy users. But the complexity of the opioid crisis – with its mix of prescription and illegal use, makes it very difficult to pick these things apart. There are some useful data sources – such as visits to emergency rooms and results of post-arrest drug tests – but this information has become harder to get hold of in recent years. We can't draw a neat graph showing the different stages of drug use like we can for the Yambuku Ebola outbreak, because the data simply aren't available. It's a common problem in outbreak analysis: things that aren't reported are by definition tough to analyse.

IN THE EARLY STAGES of a disease outbreak, there are generally two main aims: to understand transmission and to control it. These goals are closely linked. If we improve our understanding of how something is spreading, we can come up with more effective control measures. We may be able to target interventions at high-risk groups, or identify other weak links in the chain of transmission.

The relationship works the other way too: control measures

can influence our understanding of transmission. For diseases, as with drug use and gun violence, health centres often act as our windows onto the outbreak. It means that if health systems are weakened or overburdened, it can affect the quality of data coming in. During the Ebola epidemic in Liberia in August 2014, one dataset we were working with suggested that the number of new cases was leveling off in the capital Monrovia. At first this seemed like good news, but then we realised what was actually happening. The dataset was coming from a treatment unit that had reached capacity. The case reports hadn't peaked because the outbreak was slowing down; they'd stopped because the unit couldn't admit any more patients.

The interaction between understanding and control is also important in the world of crime and violence. If authorities want to know where crime is occurring, they generally have to rely on what's being reported. When it comes to using models to predict crime, this can create problems. In 2016, statistician Kristian Lum and political scientist William Isaac published an example of how reporting might influence predictions.[82] They'd focused on drug use in Oakland, California. First they'd gathered data on drug arrests in 2010, and then plugged these into the PredPol algorithm, a popular tool for predictive policing in the US. Such algorithms are essentially translation devices, taking information about an individual or location and converting it into an estimate of crime risk. According to the developers of PredPol, their algorithm uses only three pieces of data to make predictions: the type of historical crime, the place it happened and when it happened. It doesn't explicitly include any personal information – like race or gender – that could directly bias results against certain groups.

Using the PredPol algorithm, Lum and Isaac predicted where drug crimes would have been expected to occur in 2011. They also calculated the actual distribution of drug crimes that year – including those that went unreported – using data from the

National Survey on Drug Use and Health. If the algorithm's predictions were accurate, they would have expected it to flag up the areas where the crimes actually happened. But instead, it seemed to point mostly to areas where arrests had previously occurred. The pair noted that this could produce a feedback loop between understanding and controlling crime. 'Because these predictions are likely to over-represent areas that were already known to police, officers become increasingly likely to patrol these same areas and observe new criminal acts that confirm their prior beliefs regarding the distributions of criminal activity.'[83]

Some people criticised the analysis, arguing that police didn't use Predpol to predict drug crimes. However, Lum said that this is missing the wider point because the aim of predictive policing methods is to make decisions more objective. 'The implicit argument is that you want to remove human bias from the system.' If predictions reflect existing police behaviour, however, these biases will persist, hidden behind a veil of a supposedly objective algorithm. 'When you're training it with data that's generated by the same system in which minority people are more likely to be arrested for the same behaviour, you're just going to perpetuate those same issues,' she said. 'You have the same problems, but now filtered through this high-tech tool.'

Crime algorithms have more limitations than people might think. In 2013, researchers at RAND Corporation outlined four common myths about predictive policing.[84] The first was that a computer knows exactly what will happen in the future. 'These algorithms predict the risk of future events, not the events themselves,' they noted. The second myth was that a computer would do everything, from collecting relevant crime data to making appropriate recommendations. In reality, computers work best when they assist human analysis and decisions about policing, rather than replacing them entirely. The third myth was that police forces needed a high-powered model to make good predictions, whereas often the problem is getting hold of the right data. 'Sometimes you have a dataset where the information you need to make the prediction just isn't contained in that dataset,' as Lum put it.

The final, and perhaps most persistent myth, was that accurate predictions automatically lead to reductions in crime. 'Predictions, on their own, are just that – predictions,' wrote the RAND team. 'Actual decreases in crime require taking action based on those predictions.' To control crime, agencies therefore need to focus on interventions and prevention rather than simply making predictions. This is true for other outbreaks too. According to Chris Whitty, now the Chief Medical Officer for England, the best mathematical models are not necessarily the ones that try to make an accurate forecast about the future. What matters is having analysis that can reveal gaps in our understanding of a situation. 'They are generally most useful when they identify impacts of policy decisions which are not predictable by commonsense,' Whitty has suggested. 'The key is usually not that they are "right", but that they provide an unpredicted insight.'[85]

IN 2012, POLICE IN CHICAGO introduced the 'Strategic Subjects List' (SSL) to predict who might be involved in a shooting. The project was partly inspired by Andrew Papachristos's work on social networks and gun violence in the city, although Papachristos has distanced himself from the SSL.[86] The list itself is based on an algorithm that calculates risk scores for certain city inhabitants. According to its developers, the SSL does not explicitly include factors like gender, race or location. For several years, though, it wasn't clear what did go into it. After pressure from the *Chicago Sun-Times*, the Chicago Police Department finally released the SSL data in 2017. The dataset contained the information that went into the algorithm – like age, gang affiliations, and prior arrests – as well as the corresponding risk scores it produced. Researchers were positive about the move. 'It's incredibly rare – and valuable – to see the public release of the underlying data for a predictive policing system,' noted Brianna Posadas, a fellow with the social justice organisation Upturn.[87]

There were around 400,000 people in the full SSL database, with almost 290,000 of them deemed high risk. Although the algorithm didn't explicitly include race as an input, there was a noticeable difference between groups: over half of black twenty-something men in Chicago had an SSL score, compared with 6 per cent of white men. There were also a lot of people who had no clear link to violent crime, with around 90,000 'high-risk' individuals having never been arrested or a victim of crime.[88]

This raises the question of what to do with such scores. Should police monitor people who don't have any obvious connection to violence? Recall that Papachristos's network studies in Chicago focused on victims of gun violence, not perpetrators; the aim of such analysis was to help save lives. 'One of the inherent dangers of police-led initiatives is that, at some level, any such efforts will become offender-focused,' Papachristos wrote in 2016. He argued that there is a role for data in crime prevention, but it doesn't have to be solely a police matter. 'The real promise of using data analytics to identify those at risk of gunshot victimization lies not with policing, but within a broader public health approach.' He suggested that predicted victims could benefit from the support of people like social workers, psychologists, and violence interrupters.

Successful crime reduction can come in a variety of forms. In 1980, for example, West Germany made it mandatory for motor-cyclists to wear helmets. Over the next six years, motorcycle thefts fell by two thirds. The reason was simple: inconvenience. Thieves could no longer decide to steal a motorcycle on the spur of the moment. Instead, they'd have to plan ahead and carry a helmet around. A few years earlier, the Netherlands and Great

Britain had introduced similar helmet laws. Both had also seen a massive drop in thefts, showing how social norms can influence crime rates.[89]

One of the best-known ideas about how our surroundings shape crime is the 'broken windows' theory. Proposed by James Wilson and George Kelling in 1982, the idea was that small amounts of disorder – like broken windows – could spread and grow into more severe crimes. The solution, therefore, was to restore and maintain public order. The broken windows theory would become popular among police forces, most notably in New York City during the 1990s, where it inspired a heavy crackdown on minor crimes like subway fare dodging. These measures coincided with the massive drop in crime in the city, leading to claims that arrests for misdemeanours had stopped the larger offences.[90]

Not everyone was comfortable with the way that the broken windows theory was adopted. One of them was Kelling himself. He has pointed out that the original notion of broken windows was about social order rather than arrests. But the definition of public disorder can be a matter of perspective. Are those people loitering or waiting for a friend? Is that wall covered in graffiti or street art? Kelling suggested that it's not as simple as just telling police officers to restore order in an area. 'Any officer who really wants to do order maintenance has to be able to answer satisfactorily the question, "Why do you decide to arrest one person who's urinating in public and not arrest another?"' he said in 2016. 'If you can't answer that question, if you just say "Well, it's common sense," you get very, very worried.'[91]

What's more, it's not clear that aggressively punishing minor offences was the main reason for New York's decline in crime in the 90s. There's little evidence that New York's reduction was a direct result of broken windows policing. Many other US cities saw a drop in crime during that period, despite using different policing strategies. Of course, this doesn't mean broken windows policing has no effect. There's evidence that the presence of things like graffiti and stray shopping trolleys can make people far more likely to litter or use an out-of-bounds thoroughfare.[92] This suggests that minor disorder will spark other minor offences. The effect seems to work the other way too: attempts to restore order – like picking up litter – can prompt others to tidy up as well.[93] But it's quite a leap to go from such results to the conclusion that arrests for misdemeanours can explain a massive drop in violence.

So what caused the decline? Economist Steven Levitt has argued that expanded access to abortion after 1973 played a role. His theory goes that this meant there were fewer unwanted children, who would have been more likely to be involved in crime when they grew up. Others blame childhood exposure to leaded petrol and lead paint in the mid-twentieth century, which caused behavioural problems later on; when the level of exposure declined, so did crime. In fact, a recent review found that, in total, academics have proposed twenty-four different explanations for the decline in US crime during the 1990s.[94] These theories have attracted plenty of attention – as well as criticism – but the researchers involved all acknowledge that it's a complicated question. In reality, the drop in crime was likely the result of a combination of factors.[95]

This is a common problem with outbreaks that occur on long timescales. If we intervene in some way, we might have to wait a long time to see if it has an effect. In the meantime, there might be lots of other changes going on too, making it hard to measure exactly how well our intervention works. Similarly, it can be easier to focus on the immediate effects of a violent event, rather than investigate longer-term harm. Charlotte Watts has pointed out that domestic violence can be transmitted across generations, with affected children becoming involved in violence as adults. However, these children can often be forgotten when discussing interventions. 'We need to think about support for children growing up in households where there is domestic violence,' she said.

Historically, it's been difficult to analyse intergenerational transmission given the timescales involved.[96] This is where public health methods can help, suggests epidemiologist Melissa Tracy, because researchers have experience analysing long-term conditions. 'That's the strength of epidemiology, bringing that life course perspective.'

USING PUBLIC HEALTH APPROACHES to prevent crime would be hugely cost-effective, both in the US and elsewhere. Adding together the social, economic and judicial consequences of the average US murder, one study put the cost of a single killing at over $10m.[97] The problem is that the most effective solutions may not be those that people are most comfortable with. Do we want to feel like we're punishing bad people, or do we want less crime? 'When it comes to behavior change, threats and punishment are just not that effective,' said Charlie Ransford of Cure Violence. Although punishment might have some impact, Ransford suggests that other approaches generally work better. 'What is ultimately most effective at changing a person's behavior is when you try to sit down and try to listen to them and hear them out, let them air their grievances and really try to understand them,' he said. 'And then try to guide them to a healthier way of behaving.'

Projects like Cure Violence have historically focused on in-person interactions, but online social contacts are increasingly influencing the spread of violence as well. 'The environment has changed,' Ransford said. 'You need to make an adjustment. Now we're hiring workers who specialise in combing through social media to look for conflicts that need to be responded to.'

When dealing with crime and violence, it helps to understand how people are linked together. The same is true of outbreaks;

we've seen how real-life contacts can drive contagion ranging from smoking and yawning to infectious diseases and innovation. But the strength of influence online won't necessarily be the same as face-to-face encounters. 'If you think about contagion of views about acceptability of violence,' said Watts, 'the reach may be much larger, but the number of people who act might be smaller.'

It's a problem that a lot of industries are interested in. However, they generally aren't so interested in controlling contagion. When it comes to online outbreaks, people tend to care about transmission for the opposite reason. They want to make things spread.

Going viral

'YOUR NIKE ID ORDER was cancelled,' read the e-mail. It was January 2001, and Jonah Peretti was trying to get some personalised trainers. The problem was the name he'd requested; as a challenge to the company, he'd asked for his trainers to be printed with the word 'sweatshop'.[1]

Peretti, then a graduate student in the MIT Media Lab, ended up exchanging a series of e-mails with Nike. The company reiterated that it wouldn't place the order because of 'inappropriate slang'. Unable to talk them round, Peretti decided to forward the e-mail thread to a few friends. Many of them forwarded it to their friends, who forwarded it on, and on, and on. Within days, the message had spread to thousands of people. Soon the media picked up on the story too. By the end of February, the e-mail chain had gained coverage in *The Guardian* and *Wall Street Journal*, while NBC invited Peretti on to the *Today Show* to debate the issue with a Nike spokesperson. In March, the story went international, eventually reaching several European newspapers. All from that single e-mail. 'Although the press has presented my battle with Nike as a David versus Goliath parable,' Peretti later wrote, 'the real story is the battle between a company like Nike, with access to the mass media, and a network of citizens on the Internet who have only micromedia at their disposal.'[2]

The e-mail had spread remarkably far, but perhaps it had all been just a fluke? Peretti's friend and fellow PhD student Cameron Marlow seemed to think so. Marlow – who would later become head of data science at Facebook – didn't believe a person could deliberately make something take off like that. But Peretti reckoned that he could do it again. Soon after the Nike e-mail, he got a job offer from a multimedia non-profit called Eyebeam in New York. Peretti would end up leading a 'contagious media lab' at Eyebeam, experimenting with online content. He wanted to see what made things contagious and what kept them spreading.

Over the next few years, he would start to piece together features that were important for online popularity. Like how jumping on emerging news stories could drive traffic to websites. And how polarising topics got more exposure, while ever-changing content kept users coming back. His team even pioneered a 'reblog' feature that allowed people to share others' posts, a concept that would later become fundamental to how things spread on social media (just imagine how different Twitter would be without a retweet option, or Facebook without a 'share' button). Peretti would eventually move into news, helping to develop the *Huffington Post*, but those early contagion experiments stuck in his mind. Eventually, he suggested to his old boss at Eyebeam that they create a new kind of media company. One that specialised in contagion, taking their insights about popularity and applying them on a massive scale. The idea was to compile a rolling stream of viral content. They called it BuzzFeed.

NOT LONG AFTER DUNCAN WATTS published his work on small-world networks, he joined the Department of Sociology at Columbia University. During this period he became increasingly interested in online content, eventually becoming an early advisor to BuzzFeed. Although Watts had started off studying links in networks like film casts and worm brains, the world wide web contained a wealth of new data. In the early 2000s, Watts and his colleagues began to explore these online connections. In the process, they would overturn some long-held beliefs about how information spreads.

At the time, marketers were getting excited about the notion of 'influencers': everyday people who could spark social epidemics. Nowadays, the word 'influencer' has evolved to refer to everything from influential everyday people to celebrities and media personalities. But the original concept involved little-known individuals who can spark word-of-mouth outbreaks. The idea was that by targeting a few unexpectedly well-connected people, companies could get ideas to spread much further for much less cost. Rather than relying on a celebrity like Oprah Winfrey to promote their product, they could instead build enthusiasm from the ground up. 'The whole thing that made it interesting to people in the marketing world was that they could get Oprah-like impact from small budgets,' said Watts, who is now based at the University of Pennsylvania.[3]

The idea of such influencers was inspired by psychologist Stanley Milgram's famous 'small-world' experiment. In 1967, Milgram set three hundred people the task of getting a message to a specific stockbroker who lived in the town of Sharon, near Boston.[4] In the end, sixty-four of the messages would find their target. Of these, a quarter flowed through the same one person, who was a local clothing merchant. Milgram said it came as a shock to the stockbroker to find out that this merchant was apparently his biggest link to the wider world. If an innocuous merchant could be this important for the spread of a message, perhaps there were other, similarly influential people out there too?

Watts has pointed out that there are actually multiple versions of the influencer hypothesis. 'There's an interesting but not true version,' he said, 'and then there's a true but not interesting version.' The interesting version is that there are specific people – like Milgram's clothing merchant – who play a massively disproportionate role in social contagion. And if you can identify them, you can make things spread without huge marketing budgets and celebrity endorsements. It's an appealing idea, but one that doesn't hold up under scrutiny. In 2003, Watts and his colleagues at Columbia re-ran Milgram's experiment, this time with e-mails and on a much larger scale.[5] Picking eighteen different target individuals across thirteen countries, the team started almost 25,000 e-mail chains, asking each participant to get their message to a specific target. In Milgram's smaller study, the clothing merchant had appeared to be a vital link, but this wasn't the case for the e-mail chains. The messages in each chain flowed through a range of different people, rather than the same 'influencers' cropping up again and again. What's more, the Columbia researchers asked participants why they forwarded the e-mail to the people they did. Rather than sending the message to contacts who were especially popular or well connected, people tended to pick based on characteristics like location or occupation.

The experiment showed that messages don't need highly connected people to get to a specific destination. But what if we're interested simply in making something spread as far as possible? Could people who are more connected in the network – like celebrities – help ensure it takes off? A few years after the e-mail analysis, Watts and his colleagues looked at how web links propagate on Twitter. The results suggested that content was more likely to spread widely if it was posted by a person with lots of followers or a history of making things take off. Yet it was no guarantee: most of the time these people weren't successful at creating large outbreaks.[6]

Which brings us to the more basic version of the influencer hypothesis. This is simply the idea that some people can be more influential than others. There is plenty of evidence to support this. For example, in 2012 Sinan Aral and Dylan Walker studied how a person's friends influenced their choice of apps on Facebook. They found that within friendship pairings, women influenced men at a 45 per cent higher rate than they influenced other women, and over-30s were 50 per cent more influential than under-18s. They also showed that women were less susceptible to influence than men and married people were less susceptible than singles.[7]

If we want an idea to spread, we ideally need people to be both highly susceptible and highly influential. But Aral and Walker found that such people were very rare. 'Highly influential individuals tend not to be susceptible, highly susceptible individuals tend not to be influential, and almost no one is both highly influential and highly susceptible to influence,' they noted. So what effect could targeting influential people have? In a follow-up study, Aral's team simulated what would happen if the best possible people were chosen to spark a social outbreak. Compared with choosing randomly, the pair found that picking targets effectively could potentially help things spread up to twice as far. It's an improvement, but it's a long way from having a few little-known influencers who can spark a huge outbreak all by themselves.[8]

Why is it so hard to get ideas to spread from person to person? One reason is that issue of people rarely being both susceptible and influential. If someone spreads an idea to lots of susceptible people, these individuals won't necessarily pass it on much further. Then there's the structure of our interactions. Whereas financial networks are 'disassortative' – with big banks connected to lots of small ones – human social networks tend to be the opposite. From village communities to Facebook friendships, there's evidence that popular people often form social groups with other popular people.[9] It means that if we target a few popular individuals, we might get a word-of-mouth outbreak that spreads quickly, but it probably won't reach much of the network. Sparking multiple outbreaks across a network may therefore be more effective than trying to identify high profile influencers within a community.[10]

Watts has noticed that people tend to mix up the different influencer theories. They might claim to have found hidden influencers – like the merchant in Milgram's experiment – and used them to make something spread. But in reality they may have just run a mass-media campaign or paid celebrities to promote the product online, in effect bypassing word-of-mouth transmission altogether. 'People either carelessly or deliberately conflate them, to make the boring thing sound like the interesting thing,' Watts said.

The debate around influencers shows we need to think about how we are exposed to information online. Why do we adopt some ideas but not others? One reason is competition: opinions, news, and products are all fighting for our attention. A similar effect occurs with biological contagion. The pathogens behind diseases like flu and malaria are actually made up of multiple strains, which continuously compete for susceptible humans. Why doesn't one strain end up dominating everywhere? Our social behaviour probably has something to do with it. If people gather into distinct tight-knit cliques, it can allow a wider range of strains to linger in a population. In essence, each strain can find its own home territory, without having to constantly compete with others.[11] Such social interactions would also explain the huge diversity in ideas and opinions online. From political stances to conspiracy theories, social media communities frequently cluster around similar worldviews.[12] This creates the potential for 'echo chambers', in which people rarely hear views that contradict their own.

One of the most vocal online communities is the anti-vaccination movement. Members often congregate around the popular, but baseless, claim that the measles-mumps-rubella

(MMR) vaccine causes autism. The rumours started in 1998 with a scientific paper – since discredited and retracted – led by Andrew Wakefield, who was later struck off the UK medical register. Unfortunately, the British media picked up on Wakefield's claims and amplified them.[13] This led to a decline in MMR vaccination, followed by several large outbreaks of measles years later, when unvaccinated children began entering the bustling environments of schools and universities.

Despite widespread MMR rumours in the UK during the early 2000s, media reports were very different on the other side of the channel. While MMR was getting bad press in the UK, the French media were speculating about an unproven link between the hepatitis B vaccine and multiple sclerosis. More recently, there has been negative coverage of the HPV vaccine in the Japanese media, while a twenty-year-old rumour about tetanus vaccines resurfaced in Kenya.[14]

Scepticism of medicine isn't new. People have been questioning disease prevention methods for centuries. Before Edward Jenner identified a vaccine against smallpox in 1796, some would use a technique called 'variolation' to reduce their risk of disease. Developed in sixteenth-century China, variolation exposed healthy people to the dried scabs or pus of smallpox patients. The idea was to stimulate a mild form of infection, which would provide immunity to the virus. The procedure still carried a risk – around 2 per cent of variolations resulted in death – but it was much smaller than the 30 per cent chance of death that smallpox usually came with.[15]

Variolation became popular in eighteenth-century England, but was the risk worth it? French writer Voltaire observed that other Europeans thought that the English were fools and madmen to use the method. 'Fools, because they give their children the smallpox to prevent their catching it; and madmen, because they wantonly communicate a certain and dreadful distemper to their children, merely to prevent an uncertain evil.' He noted that the criticism went the other way too. 'The English, on the other side, call the rest of the Europeans cowardly and unnatural. Cowardly, because they are afraid of putting their children to a little pain; unnatural, because they expose them to die one time or other of the small-pox.'[16] (Voltaire, himself a survivor of smallpox, supported the English approach.)

In 1759, mathematician Daniel Bernoulli decided to try and settle the debate. To work out whether the risk of smallpox infection outweighed the risk from variolation, he developed the first-ever outbreak model. Based on patterns of smallpox transmission, he estimated that variolation would increase life expectancy so long as the risk of death from the procedure was below 10 per cent, which it was.[17]

For modern vaccines, the balancing act is generally far clearer. On one side, we have overwhelmingly safe, effective vaccines like MMR; on the other, we have potentially deadly infections like measles. Widespread refusal of vaccination therefore tends to be a luxury, a side effect of living in places that – thanks to vaccination – have seen little of such infections in recent decades.[18] One 2019 survey found that European countries tended to have much lower levels of trust in vaccines compared to those in Africa and Asia.[19]

Although rumours about vaccines have traditionally been country-specific, our increasing digital connectedness is changing that. Information can now spread quickly online, with automated translations helping myths about vaccination cross language barriers.[20] The resulting decline in vaccine confidence could have dire consequences for children's health. Because measles is so contagious, at least 95 per cent of a population needs to be vaccinated to have a hope of preventing outbreaks.[21] In places where anti-vaccination beliefs have spread successfully, disease outbreaks are now following. In recent years, dozens of people have died of measles in Europe, deaths that could easily have been prevented with better vaccination coverage.[22]

The emergence of such movements has drawn attention to the possibility of echo chambers online. But how much have social media algorithms actually changed our interaction with information? After all, we share beliefs with people we know in real life as well as online. Perhaps the spread of information online is just a reflection of an echo chamber that was already there?

On social media, three main factors influence what we read: whether one of our contacts shares an article; whether that content appears in our feed; and whether we click on it. According to data from Facebook, all three factors can affect our consumption of information. When the company's data science team examined political opinions among US users during 2014–2015, they found that people tended to be exposed to views that were similar to theirs, much more so than they would have been if they had picked their friends at random. Of the content that these friends posted, the Facebook algorithm – which decides what appears on users' News Feeds – filtered out another 5–8 per cent of opposing political views. And of the content people saw, they were less likely to click on articles that went against their political stance. Users were also far more likely to click on posts that appeared at the top of their feed, showing how intensely content has to compete for attention. This suggests that if echo chambers exist on Facebook, they start with our friendship choices but can then be exaggerated by the News Feed algorithm.[23]

What about the information we get from other sources? Is this similarly polarised? In 2016, researchers at Oxford University, Stanford University and Microsoft Research looked at the web browsing patterns of 50,000 Americans. They found that the articles people saw on social media and search engines were generally more polarised than the ones they came across on their favourite news websites.[24] However, social media and search engines also exposed people to a wider range of views. The stories might have had stronger ideological content, but people got to see more of the opposing side as well.

This might seem like a contradiction: if social media exposes us to a broader range of information than traditional news sources, why doesn't it help dampen the echoes? Our reaction to online information might have something to do with it. When sociologists at Duke University got US volunteers to follow Twitter accounts with opposing views, they found that people tended to retreat further back into their own political territory afterwards.[25] On average, Republicans became more conservative and Democrats more liberal. This isn't quite the same as the 'backfire effect' we saw in Chapter 3, because people weren't having specific beliefs challenged, but it does imply that reducing political polarisation isn't as simple as creating new online connections. As in real life, we may resent being exposed to views we disagree with.[26] Although having meaningful face-to-face conversations can help change attitudes – as they have with prejudice and violence – viewing opinions in an online feed won't necessarily have the same effect.

IT'S NOT JUST ONLINE CONTENT itself that can create conflict; it's also the context surrounding it. Online, we come across many ideas and communities we may not encounter much in real life. This can lead to disagreements if people post something with one audience in mind, only to have it read by another. Social media researcher danah boyd (she styles her name as lower case) calls it 'context collapse'. In real life, a chat with a close friend may have a very different tone to a conversation with a co-worker or stranger: the fact that our friends know us well means there's less potential for misinterpretation. Boyd points to events like weddings as another potential source of face-to-face context collapse. A speech that's aimed at friends could leave family uncomfortable; most of us have sat through a best man's anecdote that has made this mistake and misfired. But while weddings are (usually) carefully planned, online interactions may inadvertently include friends, family, co-work-ers, and strangers all in the same conversation. Comments can easily be taken out of context, with arguments emerging from the confusion.[27] During the COVID-19 pandemic, many disease researchers experienced this context collapse on Twitter, with a platform they'd previously used to exchange technical ideas falling under a bright public spotlight. On several occasions in 2020, I found myself widely quoted in the media after posting an offhand comment or observation, whereas six months earlier such tweets would have only been of interest to a handful of colleagues.

According to boyd, underlying contexts can also change over time, particularly as people are growing up. 'While teens' content might be public, most of it is not meant to be read by all people across all time and all space,' she wrote back in 2008. As a generation raised on social media grows older, this issue will come up more often. Viewed out of context, many historical posts – which can linger online for decades – will seem inap-propriate or ill-judged.

In some cases, people have decided to exploit the context collapse that occurs online. Although 'trolling' has become a broad term for online abuse, in early internet culture a troll was mischievous rather than hateful.[28] The aim was to provoke a sincere reaction to an implausible situation. Many of Jonah Per-etti's pre-BuzzFeed experiments used this approach, running a series of online pranks to attract attention.

Trolling has since become an effective tactic in social media debates. Unlike real life, the interactions we have online are in effect on a stage. If a troll can engineer a seemingly overblown response from their opponent, it can play well with random onlookers, who may not know the full context. The opponent, who may well have a justified point, ends up looking absurd. 'O Lord make my enemies ridiculous,' as Voltaire once said.[29]

Many trolls – of both the prankster and abuser kinds

– wouldn't behave this way in real life. Psychologists refer to it as the 'online disinhibition effect': shielded from face-to-face responses and real-life identities, people's personalities may adopt a very different form.[30] But it isn't simply a matter of a few people being trolls-in-waiting. Analysis of antisocial behaviour online has found that a whole range of people can become trolls, given the right circumstances. In particular, we are more likely to act like trolls when we are in a bad mood, or when others in the conversation are already trolling.[31]

As well as creating new types of interactions, the internet is also creating new ways to study how things spread. In the field of infectious diseases, it's generally not feasible to deliberately infect people to see how something spreads, as Ronald Ross tried to do with malaria in the 1890s. If modern researchers do run infection studies, they are usually small, expensive, and subject to careful ethical scrutiny. For the most part, we have to rely on observed data, using mathematical models to ask 'what if?' questions about outbreaks. The difference online is that it can be relatively cheap and easy to spark contagion deliberately, especially if you happen to run a social media company.

IF THEY HAD BEEN PAYING close attention, thousands of Facebook users might have noticed that on 11 January 2012, their friends were slightly happier than usual. At the same time, thousands of others may have spotted that their friends were sadder than expected. But even if they did notice a change in what their friends were posting online, it wasn't genuine change in their friends' behaviour. It was an experiment.

Researchers at Facebook and Cornell University had wanted to explore how emotions spread online, so they'd altered people's News Feeds for a week and tracked what happened. The team published the results in early 2014. By tweaking what people were exposed to, they found that emotion was contagious:

people who saw fewer positive posts had on average posted less positive content themselves, and vice versa. In hindsight, this result might seem unsurprising, but at the time it ran counter to a popular notion. Before the experiment, many people believed that seeing cheerful content on Facebook could make us feel inadequate, and hence less happy.[32]

The research itself soon sparked a lot of negative emotions, with several scientists and journalists questioning how ethical it was to run such a study. 'Facebook manipulated users' moods in secret experiment,' read one headline in the *Independent*. One prominent argument was that the team should have obtained consent, asking whether users were happy to participate in the study.[33]

Looking at how design influences people's behaviour is not necessarily unethical. Indeed, medical organisations regularly run randomised experiments to work out how to encourage healthy behaviour. For example, they might send one type of reminder about cancer screening to some people and a different one to others, and then see which gets the best response.[34] Without these kinds of experiments, it would be difficult to work out how much a particular approach actually shifted people's behaviour.

If an experiment could have a detrimental effect on users, though, researchers need to consider alternatives. In the Facebook study, the team could have waited for a 'natural experiment' – like rainy weather – to change people's emotional state, or they could have tried to answer the same research question with fewer users. Even so, it may still not have been feasible to ask for consent beforehand. In his book *Bit by Bit*, sociologist Matthew Salganik points out that psychological experiments can produce dubious results if people know what's being studied. Participants in the Facebook study might have behaved differently if they had known from the outset that the research was about emotions. If psychology researchers do deceive participants in order to get a natural reaction, however, Salganik notes that they will often debrief them afterwards.

As well as debating the ethics of the experiment, the wider research community also raised concerns about the extent of emotional contagion in the Facebook study. Not because it was big, but because it was so small. The experiment had shown that when a user saw fewer positive posts in their feed, the number of positive words in their status updates fell by an average of 0.1 per cent. Likewise, when there were fewer negative posts, negative words decreased by 0.07 per cent.

One of the quirks of huge studies is that they can flag up very small effects, which wouldn't be detectable in smaller studies. Because the Facebook study involved so many users, it was possible to identify incredibly small changes in behaviour. The study team argued that such differences were still relevant, given the size of the social network: 'In early 2013, this would have corresponded to hundreds of thousands of emotion expressions in status updates per day.' But some people remained unconvinced. 'Even if you were to accept this argument,' Salganik wrote, 'it is still not clear if an effect of this size is important regarding the more general scientific question about the spread of emotions.'

IN STUDIES OF CONTAGION, social media companies have a major advantage because they can monitor much more of the transmission process. In the Facebook emotion experiment, the researchers knew who had posted what, who had seen it, and what the effect was. External marketing companies don't have this same level of access, so instead they have to rely on alternative measurements to estimate the popularity of an idea. For example, they might track how many people click on or share a post, or how many likes and comments it receives.

What sort of ideas become popular online? In 2011,

University of Pennsylvania researchers Jonah Berger and Kath-erine Milkman looked at which *New York Times* stories people e-mailed to others. They gathered three months of data – almost 7,000 articles in total – and recorded the features of each story, as well as whether it made the 'most e-mailed' list.[35] It turned out that articles that triggered an intense emotional response were more likely to be shared. This was the case both for positive emotions, such as awe, and negative ones like anger. In con-trast, articles that evoked so-called 'deactivating' emotions like sadness were shared less often. Other researchers have found a similar emotional effect; people are more willing to spread stories that evoke feelings of disgust, for example.[36]

Yet emotions aren't the only reason we remember stories. By accounting for the emotional content of the *New York Times* articles, Berger and Milkman could explain about 7 per cent of the variation in how widely stories were shared. In other words, 93 per cent of the variation was down to something else. This is because popularity doesn't depend only on emotional content. Berger and Milkman's analysis found that having an element of surprise or practical value could also influence an article's shareability. As could the appearance of the story: an article's popularity depended on when it was posted, what section of the website it was on, and who the author was. When the pair accounted for these additional characteristics, they could explain much more of the variation in popularity.

It's tempting to think we could – in theory, at least – sift through successful and unsuccessful content to identify what makes a highly contagious tweet or article. However, even if we manage to identify features that explain why some things are more popular, these conclusions may not hold for long. Technol-ogy researcher Zeynep Tufekci has pointed to the apparent shift in people's interests as they use online platforms. On YouTube, for example, she suspected that the video recommendation algorithm might have been feeding unhealthy viewing appetites, pulling people further and further down the online rabbit hole. 'Its algorithm seems to have concluded that people are drawn to content that is more extreme than what they started with – or to incendiary content in general,' she wrote in 2018.[37] These shifting interests mean that unless new content evolves – becoming more dramatic, more evocative, more surprising – it will probably get less attention than its predecessors. Here, evolution isn't about getting an advantage; it's about survival.

The same situation arises in the biological world. Many species have to adapt simply to keep pace with their competitors. After humans came up with antibiotics to treat bacterial infections, some bacteria evolved to become resistant to common drugs. In response, we turned to even stronger antibiotics. This put pressure on bacteria to evolve further. Treatments gradually became more extreme, just to have the same impact as lesser drugs did decades earlier.[38] In biology, this arms race is known as the 'Red Queen effect', after the character in Lewis Carroll's *Through the Looking-Glass*. When Alice complains that running in the looking-glass world doesn't take her anywhere new, the Red Queen replies that, 'here, you see, it takes all the running you can do, to keep in the same place.'

This evolutionary running is about change, but it's also about transmission. Even if a new mutation crops up in bacteria, it won't automatically spread through a human population. Likewise, if new content emerges online, it's not a guarantee it will become popular. We all know of new stories and ideas that have spread widely online, but we also know of posts – perhaps including our own – that have fizzled away without notice. So how common is popularity online? What does a typical outbreak even look like?

THE RUMOURS ABOUT THE HIGGS BOSON spread gradually at first. On 1 July 2012, Twitter users started speculating that the elusive

particle – nicknamed the 'God particle'– had finally been discovered. Originally suggested by Peter Higgs in 1964, the boson
was a crucial missing piece in the subatomic jigsaw. The laws of
particle physics said it should exist, but it was yet to be observed
in reality.

That would soon change. The rumours on Twitter initially
claimed that physicists had discovered the boson at the Tevatron
particle accelerator in Illinois. The rumour outbreak grew at
a rate of about one new user per minute during this period.
The next day, researchers at the Tevatron announced that they'd
found promising – but not quite definitive – evidence that the
Higgs boson existed. The Twitter outbreak accelerated, with
more and more users joining, and attention turned to the Large
Hadron Collider at CERN. These latest rumours would prove
true: two days later, CERN researchers announced they had
indeed found the boson. As media interest in the discovery
grew, more joined the Twitter outbreak. It grew by over five
hundred users per minute for the next day or so, before peaking
soon after. By 6 July, five days after the first rumour emerged,
interest in the story had declined dramatically.[39]

When the Higgs rumours started, some users posted about
the potential discovery, while others retweeted these comments
to their own followers. If we look at how the first few hundred
of these retweets were connected, there is a huge amount of
variation in transmission (see figure on next page). Most tweets
didn't go very far, only spreading the news to one or two others.
But in the middle of the transmission network, there is a large
chain of retweets, including two large-scale transmission events,
with single users spreading the rumour to many other people.

This sort of diversity in transmission is common in online
sharing. In 2016, Duncan Watts, then based at Microsoft
Research, worked with collaborators at Stanford University to
look at 'cascades' of sharing on Twitter. The group tracked over
620 million pieces of content, noting which users had reposted

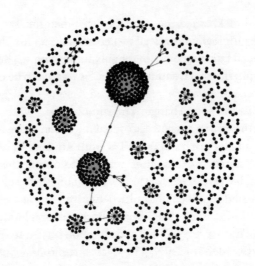

Initial retweets about the Higgs boson rumour, 1 July 2012.
Each dot represents a user, with lines showing retweets
Data: De Domenico et al., 2013

links shared by others. Some links passed between multiple
users in a long chain of transmission. Others sparked but faded
away much faster. Some didn't spread at all.[40]

For infectious diseases, we've seen there are two extreme
types of outbreaks. 'Common source' transmission occurs when
everyone gets infected from the same source, like food poisoning. At the other extreme, a propagated outbreak spreads from
person-to-person over several generations. There is a similar
diversity in online cascades. Sometimes content will spread to
lots of people from a single source – known in marketing as a
'broadcast' event – whereas on other occasions it will propagate from user to user. The Stanford and Microsoft researchers
found that broadcasts were a crucial part of the largest cascades.
About one in a thousand Twitter posts got more than 100 shares,
but only a fraction of these spread because of propagated

transmission. Of the posts that spread, there was generally a single broadcast event behind its success.

When we talk about online contagion, it's tempting to focus only on things that have become popular. However, this ignores the fact that the vast majority of things do not take off. The Microsoft team found that around 95 per cent of Twitter cascades consisted of a single tweet that nobody else shared. Of the remaining cascades, most didn't go any further than one additional step in terms of sharing. The same is true of other online platforms: it's extremely rare to get something that spreads, and even when it does, it doesn't spread beyond a few generations of transmission. Most content just isn't that contagious.[41]

IN THE PREVIOUS CHAPTER, we looked at outbreaks of shootings in Chicago, where transmission generally ended after a small number of events. Several diseases also stumble and stutter in human populations like this. For example, strains of bird flu like H5N1 and H7N9 have caused large outbreaks in poultry, but don't spread well among people (at least, not for the moment).

What sort of outbreaks should we expect if something doesn't spread very effectively? We've already looked at how we can use the reproduction number, R, to assess whether an infectious disease has the potential to spread or not; if R is above the critical value of one, there is potential for a large epidemic to occur. But even if R is below one, there's still a chance an infected person will pass the disease on to someone else. It might be unlikely, but it's possible. Unless the reproduction number is zero, we should therefore expect to get some secondary cases occasionally. These new cases may generate further generations of infection before the outbreak eventually stutters to an end.

If we know the reproduction number of a stuttering infection, can we predict how big an outbreak will be on average? It turns out that we can, thanks to a handy piece of mathematics.

As well as becoming a crucial part of outbreak analysis, it's an idea that would shape how Jonah Peretti and Duncan Watts approached viral marketing in the early days of Buzzfeed.[42]

Suppose an outbreak starts with one infectious person. By definition, this first case will generate R secondary cases on average. Then these new infections will generate R more cases each – which translates into R^2 new cases – and so on:

$$\text{Outbreak size} = 1 + R + R^2 + R^3 + \dots$$

We could try and add up all these values to work out the expected outbreak size. But fortunately there's an easier option. In the nineteenth century, mathematicians proved that there's an elegant rule we can apply to sequences like the one above. If R is between 0 and 1, the following equation is true:

$$1 + R + R^2 + R^3 + \dots = 1/(1–R)$$

In other words, as long as the reproduction number is below 1, the expected outbreak size is equal to $1/(1–R)$. Even if you're not especially interested in nineteenth-century mathematics, it's worth taking a moment to appreciate how useful this shortcut is. Rather than having to simulate how an infection might stutter along from one generation to the next until it eventually fizzles out, we can instead estimate the final outbreak size directly from the reproduction number.[43] If R is 0.8, for example, we'd expect an outbreak with $1/(1–0.8) = 5$ cases in total. And that's not all we can do. We can also work backwards to estimate the reproduction number from the average outbreak size. If outbreaks consist of five cases on average, it means R is 0.8.

In my field, we regularly use this back-of-the-envelope calculation to estimate the reproduction number of new disease threats. During the early months of 2013, there were 130 human cases of H7N9 bird flu in China. Although most of these picked up the disease from contact with poultry, there were four clusters of infection that were likely to be the result of transmission between humans.[44] Because most people didn't infect anyone else, the average size of a human H7N9 outbreak was 1.04 cases, suggesting that R in humans was a paltry 0.04.

This idea isn't only useful for diseases. During the mid-2000s, Jonah Peretti and Duncan Watts applied the same method to marketing campaigns. It meant they could get at the underlying transmissibility of an idea, rather than just describing what a campaign had looked like. In 2004, for example, anti gun violence group The Brady Campaign had sent out e-mails asking people to support new gun control measures. They encouraged recipients to forward the e-mails to their friends; some of these friends then forwarded the messages to their friends, and so on. For each e-mail that was sent, on average around 2.4 people ended up seeing the message. Based on this typical outbreak size, the reproduction number of the campaign was about 0.58. A subsequent e-mail campaign aimed to raise money for Hurricane Katrina relief efforts; this time R was 0.77. However, there wasn't always so much transmission. Spare a thought for the marketing executives trying to spread messages about cleaning products: Peretti and Watts found that e-mails promoting Tide Coldwater detergent had an R of only 0.04 (i.e. the same as H7N9 bird flu). Whereas most of the Katrina e-mails had spread between multiple people, over 99 per cent of the Tide outbreaks stuttered to an end after only one transmission event.[45]

Why do we care about measuring an infection if it won't lead to a large outbreak? For biological pathogens, a big concern is that these infections will adapt to their new hosts. During a small outbreak, viruses could pick up mutations that enable them to transmit more easily. The more people that get infected, the more chances for such adaptation. Before SARS sparked a major outbreak in Hong Kong in February 2003, there were a series of small clusters of infection in Guangdong province, in southern

China.[46] Between November 2002 and January 2003, seven out-breaks were reported in Guangdong, with between one and nine cases in each. The average outbreak size was five cases, suggesting that R may have been around 0.8 during this period. But by the time of the Hong Kong outbreak a couple of months later, SARS had a far more troubling R of more than 2.

There are several reasons the reproduction number of an infection may increase. Recall that R depends on the four DOTS: *duration* of infection, *opportunities* for transmission, *transmission* probability during each opportunity, and average *susceptibility*. For biological viruses, all of these features can influence transmission. Of the viruses that can spread among humans, the most successful tend to cause longer infections (i.e. larger duration) and spread directly from one person to another rather than via an intermediate source (i.e. more opportunities).[47] Transmission probability can also make a difference: bird flu viruses struggle to spread among people because they can't latch onto the cells in our airway as easily as human viruses can.[48]

The same sort of adaptation can happen with online content. There are many examples of online memes – such as posts and images – evolving to increase their catchiness. When Facebook researcher Lada Adamic and her colleagues analysed the spread of memes on the social network, they noticed that content would often change over time.[49] One example was a post that read: 'No one should die because they cannot afford health care and no one should go broke because they get sick.' In its original form, the meme was shared almost half a million times. But variants soon emerged, with one in every ten posts adding a mutation to the wording. Some of these edits helped the meme propagate; when people included phrases like 'post if you agree', the meme was almost twice as likely to spread. The meme was also highly resilient. After an initial peak in popularity, it persisted in one form or another for at least two years.

Even so, there seems to be a limit to the potential

contagiousness of online content. The most popular trends on Facebook during 2014–2016 all had a reproduction number of around 2. This limit seems to occur because the different components of transmission trade off against each other. Some trends – like the ice bucket challenge – involved only a few nominations per person, but came with a high probability of transmission during each nomination. Other content, such as videos and links, had far more opportunities to spread, but in reality only a few friends who saw the post reshared it.[50] Remarkably, there were no examples of Facebook content that reached lots of friends *and* had a consistently high probability of spreading to each person that saw it. This serves as a reminder of just how weak online outbreaks are compared to biological infections: even the most popular content on Facebook is ten times less contagious than measles can be.

The outlook is even worse for a typical marketing campaign. Although Jonah Peretti once bet that it was possible to get something to deliberately take off, he's since acknowledged that it's much harder to guarantee contagion when working to a client brief.[51] Consider the difference between his original Nike e-mail, which spread widely, and those later e-mail campaigns, which were far less transmissible. Peretti and Watts have pointed out that infectious diseases have millennia of evolution on their side; marketers don't have nearly as much time. 'The chances are, therefore, that even talented creatives will typically design products that exhibit R less than 1, no matter how hard they try,' they suggested.[52]

Fortunately, there is another way to increase the size of an outbreak: get the message out to more people at the start. In the above examples, we've been analysing stuttering outbreaks by assuming that one person is infectious at the start. If the reproduction number is small, this will lead to a small outbreak that fades away quickly. One way to fix this is to simply introduce more infections. Peretti and Watts call it 'big seed marketing'.

If we get a slightly contagious message to lots of people, it can pick up additional attention during subsequent small outbreaks. For example, if we send a non-contagious message to one thousand people, we'll reach one thousand people. If instead we launch a message with an R of 0.8, we'd expect to reach five thousand people in total. Much of BuzzFeed's early content became popular in this way. People saw articles on the website, then shared them with a handful of friends on sites like Facebook. Having pioneered the idea of 'reblogging' in the early 2000s, Peretti's team took full advantage of it in the decade that followed. By 2013, Buzzfeed had been named the most 'social' publisher on Facebook, with more comments, likes, and shares than any other organisation.[53] (*Huffington Post*, Peretti's former company, was second.)

If web content generally has a low R and needs multiple introductions to spread, it suggests that we shouldn't be thinking about online contagion as if it's the 1918 flu virus or SARS. Infections like pandemic flu spread easily from person to person, which means outbreaks initially grow larger and larger over several generations of transmission. In contrast, most online content won't reach many people unless there is some kind of mass broadcast event. According to Peretti, marketing companies will often talk about things going 'viral' like a disease, but they actually just mean something has become popular. 'We were thinking in terms of an actual epidemiological definition of viral, with a certain threshold of contagion that results in it growing through time,' as he once put it.[54] 'Instead of exponential decay, you get exponential growth. That is what viral is.'

Most online cascades are not viral like pandemics are; they do not grow exponentially. They are actually more like the stuttering smallpox outbreaks that occurred in Europe during the 1970s. These outbreaks would generally fade away, albeit with the occasional superspreading event leading to a large cluster of cases. Yet the smallpox superspreader analogy only goes so far, because media outlets and celebrities have a reach far beyond what's possible for biological transmission. 'A superspreader is someone who infects, like, eleven people instead of two,' Watts said. 'You don't have superspreaders who infect eleven million people.'

GIVEN THAT SOCIAL MEDIA cascades aren't the same as infectious disease outbreaks, a traditional disease model won't necessarily help us predict what will happen online. But maybe we don't need to rely on biologically inspired predictions. Given the sheer volume of data generated on social media, researchers are increasingly trying to identify transmission patterns, and use these to predict the dynamics of cascades.

How easy is it to predict online popularity? In 2016, Watts and his colleagues at Microsoft Research compiled data on almost a billion Twitter cascades.[55] They gathered data on the tweets themselves – such as the time posted and topic – as well as information about the users who initially tweeted them, such as their number of followers and whether they had a history of getting a lot of retweets. Analysing the resulting cascade sizes, they found that the content of the tweet itself provides very little information about whether it would be popular. As with their earlier analysis of influencers, the team found that a user's past tweeting success was far more important. Even so, their overall prediction ability was fairly limited. Despite having the sort of dataset a disease researcher could only dream of, the team could explain less than half the variability in cascade size.

So what explained to the other half? The researchers acknowledged that there might be some additional, as-yet-unknown features of success that could improve prediction ability. However, a large amount of the variation in popularity will depend on randomness. Even if we have detailed data about what is being tweeted and who is tweeting it, the success of a single post will inevitably depend a lot on luck. Again, this shows why it is important to spark multiple cascades, rather than trying to find a single 'perfect' tweet.

Because it's so difficult to predict a tweet's popularity before it's been posted, an alternative is to wait and look at the start of the cascade before making a prediction. This is known as the 'peeking method', because we're looking at data on the early spread before we predict what will happen next.[56] When Justin Cheng and his colleagues analysed sharing of photos on Facebook in 2014, they found that their predictions got much better once they had some data on the initial cascade dynamics. Large cascades tended to show broadcast-like spread early on, picking up lots of attention quickly. Yet the team found that some features were more elusive, even with a peeking method. 'Predicting cascade size is still much easier than predicting cascade shape,' they noted.[57]

It's not just social media content that is easier to predict after some time has passed. In 2018, Burcu Yucesoy and her colleagues at Northeastern University analysed the popularity of books on the *New York Times* bestseller list. Although it's very hard to predict whether a given book will take off in the first place, books that do become popular tend to follow a consistent pattern afterwards. The team found that most books on the bestseller list saw rapid initial growth in sales, peaking within about ten weeks of publication, which then declined to a very low level. On average, only 5 per cent of sales occurred after the first year.[58]

Despite progress in understanding online outbreaks, most analysis still relies on having good historical data. In general, it's difficult to predict the duration of a new trend ahead of time, because we don't know the underlying rules that govern transmission. However, occasionally an online cascade does follow known rules. And it was one such cascade that first sparked my interest in contagion on social media.

DRESSED IN AN 'I LOVE HATERS' baseball cap, the woman plucked the goldfish out of its bag and dropped it into a cup full of alcohol. Then she downed the drink, fish and all. A trainee lawyer, she was travelling around Australia and had performed the stunt after being nominated by a friend. The whole thing had been filmed. Before long, the video was posted on her Facebook page, along with an accompanying nomination for someone else.[59]

It was early 2014, and the woman was the latest participant in the online game of 'neknomination'. The rules were simple: players filmed themselves downing a drink, posted it on social media, then nominated others to do the same within 24 hours. The game had swept through Australia, with drinks becoming more ambitious – and alcoholic – as the nominations spread. People downed booze while skateboarding, quad biking and skydiving. Drinks varied from neat spirits to cocktails that included blended insects and even battery acid.[60]

Coverage of neknomination spread alongside the game itself. The goldfish video was widely shared, with newspapers picking up ever-more-extreme stories. When the game reached the UK, it triggered a media panic. Why was everyone doing this? How bad would it get? Should the game be banned?[61]

When neknomination hit the UK, I agreed to examine the game for a BBC radio feature.[62] I'd noticed that during games like neknomination, participants transmitted the idea to a handful of specific people, who then passed it along to others. This created a clear chain of propagated transmission, much like a disease outbreak.

If we want to predict the shape of an outbreak, there are two things we really need to know: how many additional infections each case generates on average (i.e. the reproduction number), and the lag between one round of infection and the next (i.e. the generation time). During new disease outbreaks, we rarely know these values, so we have to try and estimate them. For neknomination, though, the information was laid out as part of the game. Each person nominated 2–3 others, and these people had to do the challenge – and make their nominations – within 24 hours. When I forecast the neknomination game in 2014, I didn't have to estimate anything; I could plug the numbers straight into a simple disease model.[63]

My outbreak simulations suggested that the neknomination trend wouldn't last long. After a week or two, herd immunity would kick in, causing the outbreak to peak and begin to decline. If anything, these simple forecasts were likely to overestimate transmission. Friends tend to cluster together in real life; if multiple people nominate the same person during the game, it will reduce the reproduction number and lead to a smaller outbreak. Interest in neknomination indeed faded quickly. Despite the UK media frenzy in early February 2014, it was all but gone by the end of the month. Subsequent social media games followed a similar structure, from 'no makeup selfie' photos to the widely publicised 'ice bucket challenge'. Based on the rules of the games, my model predicted all of them would peak within a few weeks, just as they did in reality.[64]

Although nominated-based games have tended to fade away after a few weeks, social media outbreaks don't always disappear after their initial peak in popularity. Looking at popular image-based memes on Facebook, Justin Cheng and his collaborators have found that almost 60 per cent recurred at some point. On average, there was just over a month between the first and second peaks in popularity. If there were only two peaks, the second cascade of sharing was generally briefer and smaller; if there were multiple peaks, they were often a similar size.[65]

What makes a meme become popular again? The team found that a big initial peak in interest made it less likely that the meme will appear again. 'It is not the most popular cascades that recur the most,' they noted, 'but those that are only moderately popular'. This is because a small first cascade leaves more people who haven't seen the meme yet. With a large initial outbreak, there aren't enough susceptible people left to sustain transmission. For a cascade to recur, it also helps if there are several copies of the meme circulating. This is consistent with what we've already seen for stuttering outbreaks: having multiple sparks can make infections spread further.

Cheng looked at popular images, but what about other types of content? Back in 2016, I gave a public talk at London's Royal Institution. Over the next couple of years, a video of the talk somehow racked up over a million views on YouTube. Around the same time in 2016, I'd given a talk on a similar topic at Google, which had also been posted on YouTube, on a channel with a similar number of subscribers. During the same period, this one was viewed around 10,000 times. (Ideally, this popularity would have been the other way round: it turns out that if you give two related talks, but screw up a live demonstration in one of them, that's the talk that will become popular online.)

I hadn't expected the Royal Institution talk to get so much attention, but what really came as a surprise was how the views had accumulated. For its first year online, the video had gained relatively little interest, getting a hundred or so views per day. Then suddenly, in the space of a few days, it picked up more attention than it had in an entire year.

Perhaps people had started sharing it online, making it go viral? Looking at the data, the real explanation was much simpler: the video had been featured on the YouTube homepage. As the views spiked, the YouTube algorithm added it to the 'suggested video' lists that appear alongside popular videos. Almost 90 per cent of people who viewed the talk found it on the homepage or one of these lists. It was a classic broadcast event, with one source generating almost all of the views. And once the video was popular, its popularity created a feedback effect, attracting even more interest. It shows how much the video benefitted from online amplification, first by the Royal

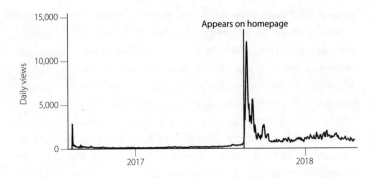

Number of YouTube views per day for my 2016 Royal Institution talk

Data: Royal Institution

Institution to get those initial few thousand views, then by the YouTube algorithm to deliver a much bigger audience.

There are three main types of popularity on YouTube. The first is where videos get a consistent, low-level number of views. This number randomly fluctuates from day-to-day, without noticeably increasing or decreasing. Around 90 per cent of YouTube videos follow this pattern. The second type of popularity is when a video suddenly gets featured on the website, perhaps in response to a news event. In this situation, almost all of the activity comes after the initial peak. The third type of popularity occurs when a video is being shared elsewhere online, gradually accumulating views before peaking and declining again. It's also possible to observe a mixture of these shapes; a shared video may get a boost by being featured then settle back down to a low level, like mine did.[66]

Video is a particularly persistent form of media, with interest tending to last much longer than for news articles. A typical social media news cycle is around two days; in the first twenty-four hours, most content comes in the form of articles, with shares and comments following afterwards.[67] However, not all news is the same. Researchers at MIT have found that false news tends to spread further and faster than true news. Maybe this is because high-profile people with lots of followers are more likely to spread falsehoods? The researchers actually found the opposite: it was generally people with fewer followers who spread the false news. If we think of contagion in terms of the four DOTS, this suggests false information spreads because the transmission probability is high, rather than there being more opportunities for spread. The reason for the high transmission probability? Novelty might have something to do with it: people like to share information that's new, and false news is generally more novel than true news.[68]

It's not just about novelty, though. To understand how things spread online, we also need to think about social reinforcement. And that means taking another look at the concept of complex contagion: sometimes we need to be exposed to an idea multiple times before we adopt it online. For example, there's evidence that we'll share memes online without much prompting, but won't share political content until we see several other people doing so. When Facebook users changed their profile picture to a '=' symbol in support of marriage equality in early 2013, on average they only did so once eight of their friends had. Complex contagion also influenced the initial adoption of many online platforms, including Facebook, Twitter and Skype.[69]

A quirk of complex contagion is that it spreads best in tight-knit communities. If people share lots of friends, it creates the multiple exposures needed for an idea to catch on. However, such ideas may then struggle to break out and spread more widely.[70] According to Damon Centola, the structure of online networks can therefore act as a barrier to complex contagion.[71] Many of our contacts online will be acquaintances rather than part of a closely linked friendship group. Whereas we might adopt a political stance if lots of our friends do, we're less likely to pick it up from a single source.

This means that complex contagion – such as nuanced political views – can have a major disadvantage on the internet. Rather than encouraging users to develop challenging, socially complex ideas, the structure of online social interactions instead favours simple, easy-to-digest content. So perhaps it's not surprising that this is what people are choosing to produce.

WITH THE RISING AVAILABILITY of data in the early twenty-first century, some suggested that researchers would no longer need to pursue explanations for human behaviour. One of them was Chris Anderson, then *Wired* editor, who in 2008 famously penned an article proclaiming the 'end of theory'. 'Who knows why people do what they do?' he wrote. 'The point is they do it, and we can track and measure it with unprecedented fidelity.'[72]

We now have vast quantities of data on human activity; it's been estimated that the amount of digital information in the world is doubling every couple of years, with much of it generated online.[73] Even so, there are a lot of things we still struggle to measure. Take those studies of obesity or smoking contagion, which show just how difficult it can be to pick apart transmission processes. Our inability to measure behaviour isn't the only problem. In a world of clicks and shares, it turns out we're not always measuring what we think we're measuring.

At first glance, clicks seem like a reasonable way to quantify interest in a story. More clicks mean more people are opening the article and potentially reading it. Surely writers who get more clicks should therefore be rewarded accordingly? Not necessarily. 'When a measure becomes a target, it ceases to be a good measure' as economist Charles Goodhart reportedly once said.[74] Rewarding success based on a simple performance metric creates a feedback loop: people start chasing the metric rather than the underlying quality it is trying to assess.

It's a problem that can occur in any field. In the run up to the

2008 financial crisis, banks paid bonuses to traders and salesmen based on their recent profits. This encouraged trading strategies that would reap benefits in the short-term, with little regard for the future. Metrics have even shaped literature. When Alexandre Dumas first wrote *The Three Musketeers* in serialised form, his publisher paid him by the line. Dumas therefore added the servant character Grimaud, who spoke in short sentences, to stretch out the text (then killed him off when the publisher said that short lines didn't count).[75]

Relying on measurements like clicks or likes can give a misleading impression of how people are truly behaving. During 2007–8, over 1.1 million people joined the 'Save Darfur' cause on Facebook, which aimed to raise money and attention in response to the conflict in Sudan. A few of the new members donated and recruited others, but most did nothing. Of the people who joined, only 28 per cent recruited someone else, and a mere 0.2 per cent donated.[76]

Despite these measurement issues, there has been a growing focus on making stories clickable and shareable. Such packaging can be highly effective. When researchers at Columbia University and the French National Institute looked at mainstream news articles mentioned by Twitter users, they found that almost 60 per cent of the links were never clicked on.[77] But this didn't stop some of the stories spreading: users shared thousands of posts featuring one of these never-clicked-on links. Evidently, many of us are happier to share something than to read it.

Perhaps it's not that surprising, given that certain types of behaviour require more effort than others. Dean Eckles, a former data scientist at Facebook, points out that it doesn't take much to get people to interact with social media in simple ways. 'That's a behaviour that's relatively easy to produce,' he said.[78] 'The behaviour we're talking about is whether your friends like or comment on the post.' Because people don't have to put much effort into performing such actions, it's much easier to get them to act. 'It's a light touch nudge for an easy to accomplish, low-cost behaviour.'

This creates a challenge for marketers. An advertising campaign might generate a lot of likes and clicks, but this isn't quite the behaviour they're interested in. They don't just want people to interact with their content; they eventually want people to buy their product or believe in their message. Just as people with more followers won't necessarily generate larger cascades, content that's more clickable or shareable won't automatically generate more revenue or advocacy.

When we're faced with a new disease outbreak, there are generally two things we want to know. What are the main routes of transmission? And which of these routes should we target to control the infection? Marketers face a similar task when designing a campaign. First, they need to know the ways someone can be exposed to a message; then they need to decide which of these routes to target. The difference, of course, is that whereas health agencies spend money to block the crucial paths of transmission, advertising agencies put money into expanding them.

Ultimately, it's a question of cost-effectiveness. Whether we're dealing with a disease outbreak or marketing campaign, we want to find the best way to allocate a limited budget. The problem is that historically it's not always been clear which path leads to which outcome. 'Half the money I spend on advertising is wasted; the trouble is I don't know which half,' as marketing pioneer John Wanamaker supposedly once said.[79]

Modern marketing has tried to tackle this problem by linking the ads people see to the actions they take afterwards. In recent years, most major websites have employed ad tracking; if companies advertise on them, they know if we saw the ads as well as whether we browsed or bought anything afterwards. Likewise, if we take an interest in their product, a company can follow us around the internet, showing us more ads.[80]

When we click on a website link, we often become the subject of a high-speed bidding war. Within about 0.03 seconds, the website server will gather all the information they have about us and send it to its ad provider. The provider then shows this information to a group of automated traders acting on behalf of advertisers. After another 0.07 seconds, the traders will have bid for the right to show us an advert. The ad provider selects the winning bid and sends the advert to our browser, which slots the advert into the webpage as it loads on the screen.[81]

People don't always realise that websites work in this way. In March 2013, the UK Labour party tweeted a link to a new press release criticising then Education Secretary Michael Gove. One Conservative MP responded by tweeting about the choice of advert on Labour's website. 'I know Labour are short of cash but having an invitation to "Date Arab girls" at top of your press release?' he wrote. Unfortunately for the MP, other users pointed out that the Labour page featured targeted advertising: the offer on display was likely to depend on a user's specific online activity.[82]

Some of the most advanced tracking has cropped up in places we might least expect it. To investigate the extent of online targeting, journalism researcher Jonathan Albright spent early 2017 visiting over a hundred extreme propaganda websites, the sort of places that are full of conspiracy theories, pseudoscience, and far-right political views. Most of the websites looked incredibly amateurish, the sort of thing a beginner would put together. But digging behind the scenes, Albright found that they concealed extremely sophisticated tracking tools. The websites were collecting detailed data on personal identity, browsing behaviour, even mouse movements. That allowed them to follow susceptible users, feeding them even more extreme content. It wasn't what users could see that made these websites so influential; it was the data harvesting that they couldn't.[83]

How much is our online data actually worth? Researchers

have estimated that users who opt-out of sharing their browsing data are worth about 60 per cent less to advertisers on Facebook. Based on Facebook's revenue in 2019, this implies that data on the behaviour of the average American user is worth at least $48 per year. Meanwhile, Google reportedly paid Apple $12bn to be the default iPhone search engine for 2019. With an estimated one billion iPhones in use, this would suggest Google value our search activity at about $12 per device.[84]

Given that our attention is so valuable, tech companies have a big incentive to keep us online. The more time we spend using their products, the more information they can collect, and the better they can tailor their content and adverts. Sean Parker, the founding president of Facebook, has previously spoken about the mindset of those who'd built early social media applications. 'That thought process was all about: "How do we consume as much of your time and conscious attention as possible?"' he said in 2016.[85] Other companies have since followed suit. 'We're competing with sleep,' joked Netflix CEO Reed Hastings in 2017.[86]

One way to keep us hooked on an app is through design. Tristan Harris, who specialises in the ethics of design, has compared the process to a magic trick. He notes that businesses will often try and guide our choices towards a specific outcome. 'Magicians do the same thing,' he once wrote. 'You make it easier for a spectator to pick the thing you want them to pick, and harder to pick the thing you don't.'[87] Magic tricks work by controlling our perception of the world; user interfaces can do the same.

Notifications are a particularly powerful way of keeping us engaged. The average iPhone user unlocks their phone over eighty times a day.[88] According to Harris, this behaviour is similar to the psychological effects of gambling addiction: 'When we pull our phone out of our pocket, we're playing a slot machine to see what notifications we got,' he suggested. Casinos capture players' attention by including payoffs that are infrequent and highly variable. Sometimes people get a reward; sometimes they get nothing. In many apps, the sender can also see if we've read their message, which encourages us to respond quicker. The more we interact with the app, the more we need to keep interacting. 'It's a social-validation feedback loop,' as Sean Parker put it. 'It's exactly the kind of thing that a hacker like myself would come up with, because you're exploiting a vulnerability in human psychology.'[89]

There are several other design features that keep us viewing and sharing content. In 2010, Facebook introduced 'infinite scrolling', removing the distraction of having to change page. Unlimited content is now common on most social media feeds; since 2015, YouTube has automatically played another video after the current one ends. Social media design is also centred on sharing; it's difficult for us to post content without seeing what others are up to.

Although not all features were originally intended to be so addictive, people are increasingly aware of how apps can influence their behaviour.[90] Even developers have become cautious of their own inventions. Justin Rosenstein and Leah Pearlman were part of the team that introduced Facebook's 'like' button. In recent years, both have reportedly tried to escape the allure of notifications. Rosenstein had his assistant put parental controls on his phone; Pearlman, who later became an illustrator, hired a social media manager to look after her Facebook page.[91]

As well as encouraging interactions, design can also hinder them. WeChat, China's vastly popular social media app, had over a billion active users in 2019. The app brings together a wide range of services: users can shop, pay bills and book travel, as well as sending messages to each other. People can also share 'Moments' (i.e. images or media) with their friends, much like the Facebook News Feed. Unlike Facebook, however, WeChat users can only ever see their friends' comments on posts.[92] This means that if you have two friends who aren't friends with each other, they can't see everything that's been said. This changes the nature of interactions. 'It prevents what I would describe as conversation from emerging,' Dean Eckles said. 'Anybody who posts anything as a comment knows that it's possible that it will be taken totally out of context, because others may see only their comment and not what happened previously in that thread.' Facebook and Twitter have widely shared posts with thousands of public comments below. In contrast, attempts at WeChat discussions inevitably look fragmented or confused, which deters users from trying.

Chinese social media discourages collective action in several ways, including deliberate barriers created by government censorship. A few years ago, political scientist Margaret Roberts and her colleagues tried to reconstruct the process of Chinese censorship. They created new accounts, posted different types of content and tracked what got removed. As they pieced together the censorship mechanisms, they discovered that criticism of leaders or policies wasn't blocked, but discussions of protests or rallies were. Roberts would later divide online censorship strategies into what she calls the 'three Fs': flooding, fear, and friction. By *flooding* online platforms with the opposing views, censors can drown out other messages. The threat of repercussions for rule breaking leads to *fear*. And removing or blocking content creates *friction* by slowing down access to information.[93]

On my first trip to mainland China, I remember trying to connect to WiFi when I arrived at my hotel. It took me a while to work out whether I was actually online. All the apps I usually might load to check my connection – Google, WhatsApp, Instagram, Twitter, Facebook, Gmail – were blocked. As well as demonstrating the power of the Chinese firewall, it made me realise how much influence US technology firms have. The bulk of my online activity is in the hands of just three companies.

We share a huge amount of information with such platforms. Perhaps the best illustration of just how much data tech companies can collect comes from a 2013 Facebook study.[94] They looked at who had typed comments on the platform but never posted them. The research team noted that the contents of the posts weren't sent back to Facebook's servers, just a record of whether someone had started typing. Maybe that was the case for this study. But regardless, it shows the level of detail with which companies can track our online behaviour and interactions. Or even, in this case, a lack of interactions.

Given the power of our social media data, organisations can have a lot to gain by accessing it. According to Carol Davidsen, who worked on the Obama campaign in the 2012 US presidential election, Facebook's privacy settings at the time made it possible to download the friendship network of everyone who'd agreed to support the campaign on the platform. These friendship connections gave the campaign a huge amount of information. 'We were actually able to ingest the entire social network of the US that's on Facebook,' she later said.[95] Facebook eventually removed this ability to gather friendship data. Davidsen claimed that, because the Republicans had been slow off the mark, the Democrats had information that their opponents didn't have. Such data analysis didn't break any rules, but the experience raised questions about how information is collected and who has control of it. 'Who owns the fact that you and I are friends?' as Davidsen put it.

At the time, many hailed the Obama campaign's use of data as innovative.[96] It was a modern method for a new political era. Just as the finance industry had got excited about new mortgage products in the 1990s, social media was seen as something that would change politics for the better. But much like those financial products, it wasn't an attitude that would last.

'HEY LOVELY YOU GONNA VOTE in the election? & for who?' In the run up to the 2017 UK general election, thousands of people looking for a date on the Tinder app got a political chat-up line instead. Londoners Charlotte Goodman and Yara Rodrigues Fowler had wanted to encourage their fellow twenty-somethings to vote for Labour, so designed a chatbot to reach a wide audience.

Once a volunteer installed the bot, it automatically set their Tinder location to somewhere in a marginal constituency, swiped 'yes' to every person, and started chatting to any matches. If the initial message was well received, volunteers could take over and start talking for real. The bot sent over 30,000 messages in total, reaching people who canvassers might not usually talk to. 'The occasional match was disappointed to be talking to a bot instead of a human, but there was very little negative feedback,' Goodman and Rodrigues Fowler later wrote. 'Tinder is too casual a platform for users to feel hoodwinked by some political conversation.'[97]

Bots make it possible to have a vast number of interactions at the same time. With a linked network of bots, people can perform actions at a scale that simply wouldn't be feasible if a human had to do it all manually. These botnets can consist of thousands, if not millions of accounts. Like human users, these bots can post content, start conversations, and promote ideas. However, the role of such accounts has come under scrutiny in recent years. In 2016, two votes shook the Western world: in June, Britain voted to leave the EU; in November, Donald Trump won the US presidency. What had caused these events? In the aftermath, speculation grew that false information – much of it created by Russia and far-right groups – had been spread widely during these elections. Vast numbers of people in the UK, and then vast numbers in the US, had been duped by fake stories posted by bots and other questionable accounts.

At first glance, the data seem to support this story. There's evidence that over 100 million Americans may have seen Facebook posts backed by Russia during the 2016 election. And on

Twitter, almost 700,000 people in the US were exposed to Russian-linked propaganda, spread by 50,000 bot accounts.[98] The idea that many voters fell for propaganda posted by fake websites and foreign spies is an appealing narrative, especially for those of us who were politically opposed to Brexit and Trump. But if we look more closely at the evidence, this simple story starts to fall apart.

Despite Russia-linked propaganda circulating during the 2016 US election, Duncan Watts and David Rothschild have pointed out that a lot of other content was as well. Facebook users may have been exposed to Russian content, but during that period American users saw over 11 *trillion* posts on the platform. For every Russian post people were exposed to, on average there were almost 90,000 other pieces of content. Meanwhile on Twitter, less than 0.75 per cent of election-related tweets came from accounts linked with Russia. 'In sheer numerical terms, the information to which voters were exposed during the election campaign was overwhelmingly produced not by fake news sites or even by alt-right media sources, but by household names,' noted Watts and Rothschild.[99] Indeed, it's been estimated that in the first year of his campaign, Trump gained almost $2bn worth of free mainstream media coverage.[100] The pair highlighted the media focus on the Hillary Clinton email controversy as one example of what outlets chose to inform their readers about. 'In just six days, the *New York Times* ran as many cover stories about Hillary Clinton's emails as they did about all policy issues combined in the 69 days leading up to the election.'

Other researchers have reached a similar conclusion about the scale of false news sources in 2016. Brendan Nyhan and his colleagues found that although some US voters consumed a lot of news from dubious websites, these people were in the minority. On average, only 3 per cent of the articles that people viewed were published by websites peddling false stories. They later published a follow-up analysis of the 2018 midterms; the results suggested that dodgy news had an even smaller reach during this election. In the UK, there was also little evidence of Russian content dominating conversations on Twitter or YouTube in the run up to the EU referendum.[101]

This might seem to suggest that we shouldn't be concerned about bots and questionable websites, but again it's not quite that simple. When it comes to online manipulation, it turns out that something much subtler – and far more troubling – has been happening.

BENITO MUSSOLINI ONCE SAID 'it is better to live one day as a lion than 100 years as a sheep'. According to the Twitter user @ilduce2016, though, the quote actually comes from Donald Trump. Originally created by a pair of journalists at *Gawker*, this Twitter bot has sent thousands of tweets misattributing Mussolini lines to Trump. Eventually one of the posts caught Trump's attention: on 28 February 2016, just after the fourth Republican primary, he retweeted the lion quote.[102]

Whereas some social media bots target a mass audience, others have a much narrower range. Known as 'honey pot bots', they aim to attract the attention of specific users and lure them into responding.[103] Remember how Twitter cascades often rely on a single 'broadcast' event? If you want to get a message to spread, it helps if someone high profile can amplify it for you. Because many outbreaks won't spark, it also helps to have a bot that can repeatedly try: @ilduce2016 posted over two thousand times before Trump finally retweeted a quote. Bot creators seem to be aware of how powerful this approach can be. When Twitter bots posted dubious content during 2016–17, they disproportionately targeted popular users.[104]

It's not just bots that use this targeting strategy. Following the 2018 shooting at Marjory Stoneman Douglas High School in Parkland, Florida, there were reports that the shooter had been a member of a small white supremacist group based in the state capital Tallahassee. However, the story was a hoax. It had started with trolls on online forums, who'd managed to persuade curious reporters that it was a genuine claim. 'All it takes is a single article,' noted one user. 'And everyone else picks up the story.'[105]

Although researchers like Watts and Nyhan have suggested that people didn't get much of their information from dubious online sources in 2016, it doesn't mean it's not a problem. 'I think it really matters, but it doesn't quite matter in the way that people think it does,' said Watts. When fringe groups post false ideas or stories on Twitter, they aren't necessarily trying to reach mass audiences. Not initially, at least. Instead, they are often targeting those journalists or politicians who spend a lot of time on social media. The hope is that these people will pick up on the idea and spread it to a wider audience. During 2017, for instance, journalists regularly quoted messages from a Twitter user named @wokeluisa, who appeared to be a young political science graduate from New York. In reality, though, the account was run by a Russian troll group, who were apparently targeting media outlets to build credibility and get messages amplified.[106] This is a common tactic among groups who want ideas to spread. 'Journalists aren't just part of the game of media manipulation,' suggested Whitney Phillips, who researches online media at Syracuse University. 'They're the trophy.'[107]

Once a media outlet picks up on a story, it can trigger a feedback effect, with others covering it too. A few years ago, I inadvertently experienced this media feedback first hand. It started when I tipped off a journalist at *The Times* about a mathematical quirk in the new National Lottery (at the time, I'd just written a book about the science of betting). Two days later the story appeared in print. The morning it was published, I got an 8.30am message from a producer at ITV's *This Morning*, who'd seen the story. By 10.30am, I was live on national television.

Soon after, I received a message from BBC *Radio 4*; they'd also read the article, and wanted to get me on their flagship lunchtime show. More coverage would follow. I'd end up reaching an audience of millions, all from that one initial story.

My experience was a harmless, if surreal, accident. But others have made a strategic effort to exploit media feedback effects. This is how false information can spread widely, despite the fact that most of the public avoid fringe websites. In essence, it's a form of information laundering. Just as drug cartels might funnel their money through legitimate businesses to hide its origins, online manipulators will get credible sources to amplify and spread their message, so the wider population will hear the idea from a familiar personality or outlet rather than an anonymous account.

Such laundering makes it possible to influence debate and coverage surrounding an issue. With careful targeting and amplification, manipulators can create the illusion of widespread popularity for specific policies or political candidates. In marketing, this strategy is known as 'astroturfing', because it artificially mimics grassroots support. This makes it harder for journalists and politicians to ignore the story, so eventually it becomes real news.

Of course, media influence isn't a recent development; it's long been known that journalists can shape the news cycle. When Evelyn Waugh wrote his 1938 satirical novel *Scoop*, he included a tale about a star reporter named Wenlock Jakes, who is sent to cover a revolution. Unfortunately, Jakes oversleeps on his train and wakes up in the wrong country. Not realising his mistake, he makes up a story about 'barricades in the streets, flaming churches, machine guns answering the rattle of his typewriter'. Other journalists, not wanting to be left out, arrive and concoct similar stories. Before long, stocks plummet and the country suffers an economic crash, leading to a state of emergency and finally a revolution.

Waugh's tale was fictional, but the underlying news feedback he describes still occurs. However, there are some major differences with modern information. One is the speed with which it can spread. Within hours, something can grow from a fringe meme into a mainstream talking point.[108] Another difference is the cost of producing contagion. Bots and fake accounts are fairly cheap to create, and mass amplification by politicians or news sources is essentially free. In some cases, popular false articles can even make money by bringing in advertising revenue. Then there's the potential for 'algorithmic manipulation': if a group can use fake accounts to manufacture the sort of reactions that are valued by social media algorithms – such as comments and likes – they may be able to get a topic trending even if few people are actually talking about it.

Given these new tools, what sort of things have people tried to make popular? Since 2016, 'fake news' has become a common term to describe manipulative online information. However, it's not a particularly helpful phrase. Technology researcher Renée DiResta has pointed out that 'fake news' can actually refer to several different types of content, including clickbait, conspiracy theories, misinformation, and disinformation. As we've seen, clickbait simply tries to entice people to visit a page; the links will often lead to real news articles. In contrast, conspiracy theories tweak real-life stories to include a 'secret truth', which may become more exaggerated or elaborate as the theory grows. Then we have misinformation, which DiResta defines as false content that is generally shared by accident. This can include hoaxes and practical jokes, which are created to be deliberately false but are then inadvertently spread by people who believe them to be true.

Finally, we have the most dangerous form of fake news: disinformation. A common view of disinformation is that it's there to make you believe something false. However, the reality is subtler than this. When the KGB trained their foreign agents during the Cold War, they taught them how to create contradictions in public opinion and undermine confidence in accurate news.[109] This is what disinformation means. It's not there to persuade you that false stories are true, but to make you doubt the very notion of truth. The aim is to shift facts around, making the reality difficult to pin down. And the KGB wasn't just good at seeding disinformation; they knew how to get it amplified. 'In the quaint old days when KGB spies deployed the tactic, the goal was pickup by a major media property,' as DiResta put it, 'because that provided legitimization and took care of distribution.'[110]

In the past decade or so, a handful of online communities have been particularly successful at getting their messages picked up. One early example emerged in September 2008, when a user posted on the Oprah Winfrey Show's online message board. The user claimed to represent a massive paedophile network, with over 9,000 members. But the post wasn't quite what it seemed: the phrase 'over 9,000' – a reference to a fighter shouting about their opponent's power level in the cartoon *Dragon Ball Z* – was actually a meme from 4chan, an anonymous online message board popular with trolls. To the delight of 4chan users, Winfrey took the paedophilia claim seriously and read out the phrase on air.[111]

Online forums like 4chan – and others such as Reddit and Gab – in effect act as incubators for contagious memes. When users post images and slogans, it can spark large numbers of new variants. These newly mutated memes spread and compete on the forums, with the most contagious ones surviving and the weaker ones disappearing. It's a case of 'survival of the fittest', the same sort of process that occurs in biological evolution.[112] Although it isn't anything like the millennia-long timescales that pathogens have had, this crowd-sourced evolution can still give online content a major advantage.

One of the most successful evolutionary tricks honed by trolls has been to make memes absurd or extreme, so it's unclear whether they are serious or not. This veneer of irony can help unpleasant views spread further than they would otherwise. If users take offence, the creator of the meme can claim it was a joke; if users assume it was a joke, the meme goes uncriticised. White supremacist groups have also adopted this tactic. A leaked style guide for the *Daily Stormer* website advised its writers to keep things light to avoid putting off readers: 'generally, when using racial slurs, it should come across as half-joking.'[113]

As memes rise in prominence, they can become an effective resource for media-savvy politicians. In October 2018, Donald Trump adopted the slogan 'Jobs Not Mobs', claiming that Republicans favoured the economy over immigration. When journalists traced the idea to its source, they found that the meme had probably originated on Twitter. It had then spent time evolving on Reddit forums, becoming catchier in the process, before spreading more widely.[114]

It's not just politicians who can pick up on fringe content. Online rumours and misinformation have spurred attacks on minority groups in Sri Lanka and Myanmar, as well as outbreaks of violence in Mexico and India. At the same time, disinformation campaigns have worked to stir up both sides of a dispute. During 2016 and 2017, Russian troll groups reportedly created multiple Facebook events, with the aim of getting opposing crowds to organise far-right protests and counter-protests.[115] Disinformation around specific topics like vaccination can also feed into wider social unrest; mistrust of science tends to be associated with mistrust in government and the justice system.[116]

The spread of harmful information is not a new problem. Even the term 'fake news' has emerged before, briefly becoming popular in the late 1930s.[117] But the structure of online networks has made the issue faster, larger and less intuitive. Like certain infectious diseases, information can also evolve to spread more efficiently. So what can we do about it?

THE GREAT EAST JAPAN EARTHQUAKE was the largest in the country's history. It was powerful enough to shift the Earth on its axis by several inches, with forty-metre-high tsunami waves following soon after. Then the rumours started. Three hours after the earthquake hit on 11 March 2011, a Twitter user claimed that poisonous rain might fall because a gas tank had exploded. The explosion had been real, but the dangerous rain wasn't. Still, it didn't stop the rumours. Within a day, thousands of people had seen and shared the false warning.[118]

In response to the rumour, the government in the nearby city of Urayasu tweeted a correction. Despite the false information having a head start, the correction soon caught up. By the following evening, more users had retweeted the correction than the original rumour. According to a group of Toyko-based researchers, a quicker response could have been even more successful. Using mathematical models, they estimated that if the correction had been issued just two hours earlier, the rumour outbreak would have been 25 per cent smaller.

Prompt corrections might not stop an outbreak, but they can slow it down. Researchers at Facebook have found that if users are quick to point out that their friend has shared a hoax – such as a get-rich-quick scheme – there's an up to 20 per cent chance the friend will delete the post.[119] In some cases, companies have deliberately slowed down transmission by altering the structure of their app. After a series of attacks in India linked to false rumours, WhatsApp made it harder for users to forward content. Rather than being able to share messages with over a hundred people, users in India would be limited to just five.[120]

Notice how these counter-measures work by targeting different aspects of the reproduction number. WhatsApp reduced the opportunities for transmission. Facebook users persuaded their friends to remove a post, which reduced the duration of infectiousness. Urayasu City Hall reduced susceptibility, by exposing thousands of people to the correct information before they saw the rumour. As with diseases, some parts of the reproduction number may be easier to target than others. In 2019, Pinterest announced they'd blocked anti-vaccination content from appearing in searches (i.e. removing opportunities for transmission), having struggled to remove it completely, which would have curbed the duration of infectiousness. [121]

Then there's the final aspect of the reproduction number: the inherent transmissibility of an idea. Recall how there are media guidelines for reporting events like suicides, to limit the potential for contagion effects. Researchers like Whitney Phillips have suggested we treat manipulative information in the same way, avoiding coverage that spreads the problem further. 'As soon as you're reporting on a particular hoax or some other media manipulation effort, you're legitimising it,' she said, 'and you're essentially providing a blueprint for what somebody down the road knows is going to work.'[122]

Recent events have shown that some media outlets still have a long way to go. In the aftermath of the 2019 mosque shootings in Christchurch, New Zealand, several outlets ignored well-established guidelines for reporting on terrorist attacks. Many published the shooter's name, detailed his ideology, or even displayed his video and linked to his manifesto. Worryingly, this information caught on: the stories that were widely shared on Facebook were far more likely to have broken reporting guidelines.[123]

This shows we need to rethink about how we interact with malicious ideas, and who is really benefitting when we give them our attention. A common argument for featuring extreme views is that they would spread anyway, even without media amplification. But studies of online contagion have found the opposite: content rarely goes far without broadcast events to amplify it. If an idea becomes popular, it's generally because well-known personalities and media outlets have helped it spread, whether deliberately or inadvertently.

Unfortunately, the changing nature of journalism has made it harder to resist media manipulators. An increasing desire for online shares and clicks has left many outlets open to exploitation by people who can deliver contagious ideas, and the attention that comes with them. That attracts trolls and manipulators, who have a much better understanding of online contagion than most. From a technological point of view, most manipulators aren't abusing the system. They're following its incentives. 'What's insidious about it is that they use social media in precisely the ways it was designed to be used,' Phillips said. In her research, she has interviewed dozens of journalists, many of whom felt uneasy knowing they are profiting from stories about extreme views. 'It's really good for me, but really bad for the country,' one reporter told her. To reduce the potential for contagion, Phillips argues that the manipulation process needs to be discussed alongside the story. 'Making clear in the reporting that the story itself is part of an amplification chain, that the journalist is part of an amplification chain, that the reader is part of an amplification chain – these things need to be really foregrounded in coverage.'

Although journalists can play a large role in outbreaks of information, there are other links in the transmission chain too, most notably social media platforms. But studying contagion on these platforms is not as straightforward as reconstructing a sequence of disease cases or gun incidents. The online ecosystem has a massive number of dimensions, with trillions of social interactions and a huge array of potential transmission routes. Despite this complexity, though, proposed solutions to harmful information are often one-dimensional, with suggestions that we need to do more of something or less of something.

As with any complex social question, there's unlikely to be a simple, definitive answer. 'I think the shift we're going through is akin to what happened in the United States on the war on drugs,' said Brendan Nyhan.[124] 'We're moving from "this is a problem that we have to solve" to "this is a chronic condition we have to manage". The psychological vulnerabilities that make humans prone to misperceptions aren't going to go away. The online tools that help it circulate aren't going to go away.'

What we can do, though, is try and make media outlets, political organisations, and social media platforms – not to mention ourselves – more resistant to manipulation. To start with, that means having a much better understanding of the transmission process. It's not enough to concentrate on a few groups, or countries, or platforms. Like disease outbreaks, information rarely respects boundaries. Just as the 1918 'Spanish flu' was blamed on Spain because it was the only country reporting cases, our picture of online contagion can be skewed by where we see outbreaks. In recent years, researchers have published almost five times more studies looking at contagion on Twitter than on Facebook, despite the latter having seven times more users.[125] This is because, historically, it's been much easier for researchers to access public Twitter data than to see what's spreading on closed apps like Facebook or WhatsApp.

There's hope the situation could change – in 2019, Facebook announced it was partnering with twelve teams of academics to study the platform's effect on democracy – but we still have a long way to go to understand the wider information eco-system.[126] One of the reasons online contagion is so hard to investigate is that it's been difficult for most of us to see what other people are actually exposed to. A couple of decades ago, if we wanted to see what campaigns were out there, we could pick up a newspaper or turn on our televisions. The messages themselves were visible, even if their impact was unclear. In outbreak terms, everyone could see the sources of infection, but nobody really understood how much transmission was happening, or which infection came from which source. Contrast this with the rise of social media, and manipulation campaigns that follow specific users around the internet. When it comes to spreading ideas, groups seeding information in recent years have had a much better idea about the paths of transmission, but the sources of infection have been invisible to everyone else.[127]

Uncovering and measuring the spread of misinformation and disinformation will be crucial if we want to design effective counter-measures. Without a good understanding of contagion, there's a risk of either blaming the wrong source, 'bad air'-style, or proposing simplistic strategies like abstinence, which – as with STI prevention – might work in theory but not in practice. By accounting for the transmission process, we'll have a better chance of avoiding epidemiological errors like these.

We'll also be able to take advantage of knock-on benefits. When something is contagious, a control measure will have both a direct and indirect effect. Think about vaccination. Vaccinating someone has a direct effect because they now won't get infected; it also has an indirect effect because they won't pass an infection on to others. When we vaccinate a population, we therefore benefit from both the direct and indirect effects.

The same is true of online contagion. Tackling harmful content will have a direct effect – preventing a person from seeing it – as well as an indirect effect, preventing them spreading it to others. This means well-designed measures may prove disproportionately effective. A small drop in the reproduction number can lead to a big reduction in the size of an outbreak.

'IS SPENDING TIME ON SOCIAL MEDIA bad for us?' asked two Facebook researchers in late 2017. David Ginsburg and Moira Burke had weighed up the evidence about how social media use affects wellbeing. The results, published by Facebook, suggested that not all interactions were beneficial. For example, Burke's research had previously found that receiving genuine messages from close friends seemed to improve users' wellbeing, but receiving casual feedback – such as likes – did not. 'Just like in person, interacting with people you care about can be beneficial,' Ginsburg and Burke suggested, 'while simply watching others from the sidelines may make you feel worse.'[128]

The ability to test common theories about human behaviour is a big advantage of online studies. In the past decade or so, researchers have used massive datasets to question long-standing ideas about the spread of information. This research has already challenged misconceptions about online influence, popularity, and success. It's even overturned the very concept of something 'going viral'. Online methods are also finding their way back into disease analysis; by adapting techniques used to study online memes, malaria researchers have found new ways to track the spread of disease in Central America.[129]

Social media might be the most prominent way our interactions have changed, but it's not the only network that's been growing in our lives. As we shall see in the next chapter, technological connections are expanding in other ways, with new links permeating through our daily routines. Such technology can be hugely beneficial, but it can also create new risks. In the world of outbreaks, every new connection is a potential new route of contagion.

6

How to own the internet

WHEN A MAJOR CYBER-ATTACK took down websites including
Netflix, Amazon, and Twitter, the attackers included kettles,
fridges, and toasters. During 2016, a piece of software called
'Mirai' had infected thousands of smart household devices
worldwide. These items increasingly allow users to control
things like temperature via online apps, creating connections
that are vulnerable to infection. Once infected with Mirai, the
devices had formed a vast network of bots, creating a powerful
online weapon.[1]

On 21 October that year, the world discovered that the weapon
had been fired. The hackers behind the botnet had chosen to
target Dyn, a popular domain name system. These systems
are crucial for navigating the web. They convert familiar web
addresses – like Amazon.com – into a numeric IP address that
tells your computer where to find the site on the web. Think of
it like a phonebook for websites. The Mirai bots attacked Dyn
by flooding it with unnecessary requests, bringing the system
to a halt. Because Dyn provides details for several high profile
websites, it meant people's computers no longer knew how to
access them.

Systems like Dyn handle a lot of requests every day without
problems, so it takes a massive effort to overwhelm them. That

effort came from the sheer scale of the Mirai network. Mirai was able to pull off its attack – one of the largest in history – because the software wasn't infecting the usual culprits. Traditionally, botnets have consisted of computers or internet routers, but Mirai had spread through the 'internet of things'; as well as kitchenware, it had infected devices like smart TVs and baby monitors. These items have a clear advantage when it comes to organising mass cyber-attacks: people turn off their computers at night, but often leave other electronics on. 'Mirai was an insane amount of firepower,' one FBI agent later told *Wired* magazine.[2]

The scale of the Mirai attack showed just how easily artificial infections can spread. Another high-profile example would emerge a few months later, on 12 May 2017, when a piece of software called 'WannaCry' started holding thousands of computers to ransom. First it locked users out of their files, then displayed a message telling users they had three days to transfer $300 worth of Bitcoin to an anonymous account. If people refused to pay up, their files would be permanently locked. WannaCry would end up causing widespread disruption. When it hit the computers of the UK National Health Service, it resulted in the cancellation of 19,000 appointments. In a matter of days, over a hundred countries would be affected, leading to over $1bn worth of damage.[3]

Unlike outbreaks of social contagion or biological infections, which may take days or weeks to grow, artificial infections can operate on much faster timescales. Outbreaks of malicious software – or 'malware' for short – can spread widely within a matter of hours. In their early stages, the Mirai and WannaCry outbreaks were both doubling in size every 80 minutes. Other malware can spread even faster, with some outbreaks doubling in a matter of seconds.[4] However, computational contagion hasn't always been so rapid.

THE FIRST EVER COMPUTER VIRUS to spread 'in the wild' outside of a laboratory network started as a practical joke. In February 1982, Rich Skrenta wrote a virus that targeted Apple II home computers. A fifteen-year-old high school student in Pennsylvania, Skrenta had designed the virus to be annoying rather than harmful. Infected machines would occasionally display a short poem he'd written.[5]

The virus, which he called 'Elk Cloner', spread when people swapped games between computers. According to network scientist Alessandro Vespignani, most early computers weren't networked, so computer viruses were much like biological infections. 'They were spreading on floppy disks. It was a matter of contact patterns and social networks.'[6] This transmission process meant that Elk Cloner didn't get much further than Skrenta's wider friendship group. Although it reached his cousins in Baltimore and made its way onto the computer of a friend in the US Navy, these longer journeys were rare.

Yet the era of localised, relatively harmless viruses wouldn't last long. 'Computer viruses quickly drifted into a completely different world,' said Vespignani. 'They were mutating. The transmission routes were different.' Rather than relying on human interactions, malware adapted to spread directly from machine to machine. As malware became more common, the new threats needed some new terminology. In 1984, computer scientist Fred Cohen came up with the first definition of a computer virus, describing it as a program that replicates by infecting other programs, just as a biological virus needs to infect host cells to reproduce.[7] Continuing the biological analogy, Cohen contrasted viruses with 'computer worms', which could multiply and spread without latching onto other programs.

Online worms first came to public attention in 1988 thanks to the 'Morris worm', created by Cornell student Robert Morris. Released on 2 November, it spread quickly through ARPANET, an early version of the Internet. Morris claimed that the worm was meant to transmit silently, in an effort to estimate the size of the network. But a small tweak in its code would cause some big problems.

Morris had originally coded the program so that when it reached a new computer, it would start by checking whether the machine was already infected, to avoid installing multiple worms. The problem with this approach is that it made it easy for users to block the worm; they could in essence 'vaccinate' their computer against it by mimicking an infection. To get around this issue, Morris had the worm sometimes duplicate itself on a machine that was already infected. But he underestimated the effect this would have. When it was released, the worm spread and replicated far too quickly, causing many machines to crash.[8]

The story goes that the Morris worm eventually infected 6,000 computers, around 10 per cent of the internet at the time. According to Morris's contemporary Paul Graham, however, this was just a guess, which soon spread. 'People like numbers,' he later recalled. 'And so this one is now replicated all over the Internet, like a little worm of its own.'[9]

EVEN IF THE MORRIS OUTBREAK NUMBER were true, it would pale in comparison to modern malware. Within a day of the Mirai outbreak starting in August 2016, almost 65,000 devices had been infected. At its peak, the resulting botnet consisted of over half a million machines, before shrinking in size in early 2017.

Yet Mirai did share a similarity with the Morris worm, in that its creators hadn't expected the outbreak to get so out of hand. Although Mirai would hit headlines when it affected websites like Amazon and Netflix in October 2016, the botnet was initially designed for a more niche reason. When the FBI traced its origins, they discovered it had started with a twenty-one-year-old college student named Paras Jha, his two friends, and the computer game Minecraft.

Minecraft has over fifty million active users globally, who play together in vast online worlds. The game has been hugely profitable for its creator, who bought a $70m mansion after selling Minecraft to Microsoft in 2014.[10] It has also been lucrative for people who run the independent servers that host Minecraft's different virtual landscapes. While most online multiplayer games are controlled by a central organisation, Minecraft operates as a free market: people can pay to access whichever server they want. As the game became more popular, some server owners found themselves making hundreds of thousands of dollars a year.[11]

Given the increasing amount of money on the line, a few owners decided to try and take out their rivals. If they could direct enough fake activity at another server – what's known as a 'distributed denial of service' (DDoS) attack – it would slow down the connection for anyone playing. This would frustrate users into looking for an alternative server, ideally the one owned by the people who organised the attack. An online arms market emerged, with mercenaries selling increasingly sophisticated DDoS attacks, and in many cases also selling protection against them.

This was where Mirai came in. The botnet was so powerful it would be able to outcompete any rivals attempting to do the same thing. But Mirai didn't remain in the Minecraft world for long. On 30 September 2016, a few weeks before the Dyn attack, Jha and his friends published the source code behind Mirai on an internet forum. This is a common tactic used by hackers: if code is publicly available, it's harder for authorities to pin down its creators. Someone else – it's not clear who – then downloaded the trio's code and used it to target Dyn with a DDoS attack.

Mirai's original creators – who were based in New Jersey, Pittsburgh and New Orleans – were eventually caught after the FBI seized infected devices and painstakingly followed the chain of transmission back to its source. In December 2017, the three pleaded guilty to developing the botnet. As part of their sentence, they agreed to work with the FBI to prevent other similar attacks in the future. A New Jersey court also ordered Jha to pay $8.6 million in restitution.[12]

The Mirai botnet managed to bring the internet to a halt by targeting the Dyn web address directory, but on other occasions, web address systems have helped someone stop an attack. As the WannaCry outbreak was growing in May 2017, British cybersecurity researcher Marcus Hutchins got hold of the worm's underlying code. It contained a lengthy gibberish web address – iuqerfsodp9ifjaposdfjhgosurijfaewrwergwea.com – that WannaCry was apparently trying to access. Hutchins noticed the domain wasn't registered, so bought it for $10.69. In doing so, he inadvertently triggered a 'kill switch' that ended the attack. 'I will confess that I was unaware registering the domain would stop the malware until after I registered it, so initially it was accidental,' he later tweeted.[13] 'So I can only add "accidentally stopped an international cyber attack" to my résumé.'

One of the reasons Mirai and WannaCry spread so widely is that the worms were very efficient at finding vulnerable machines. In outbreak terms, modern malware can create a lot of opportunities for transmission, far more than their predecessors were capable of. In 2002, computer scientist Stuart Staniford and his colleagues wrote a paper titled 'How to own the Internet in Your Spare Time'[14] (in hacker culture, 'own' means 'control completely'). The team showed that the 'Code Red' worm, which had spread through computers the previous year, had actually been fairly slow. On average, each infected server had only infected 1.8 other machines per hour. This was still much faster than measles, one of the most contagious human infections: in a susceptible population, a person who has measles will infect 0.1 others per hour on average.[15] But it was still slow enough to mean that, like a human outbreak, Code Red took a while to really take off.

Staniford and his co-authors suggested that, with a more streamlined, efficient worm, it would be possible to get a much faster outbreak. Borrowing from Andy Warhol's famous 'fifteen minutes of fame' quote, they called this hypothetical creation a 'Warhol worm', because it would be able to reach most of its targets within this time. However, the idea didn't stay hypothetical for long. The following year, the world's first Warhol worm surfaced when a piece of malware called 'Slammer' infected over 75,000 machines.[16] Whereas the Code Red outbreak had initially doubled in size every 37 minutes, Slammer doubled every 8.5 *seconds*.

Slammer had spread quickly at first, but it soon burned itself out as it became harder to find susceptible machines. The eventual damage was also limited. Although the sheer volume of Slammer infections slowed down many servers, the worm wasn't designed to harm the machines it infected. It's another example of how malware can come with a range of symptoms, just like real-life infections. Some worms are near invisible or display poems; others hold machines to ransom or launch DDoS attacks.

As shown by the Minecraft server attacks, there can be an active market for the most powerful worms. Such malware is commonly sold in hidden online marketplaces, like the 'dark net' markets that operate outside the familiar, visible websites we can access with regular search engines. When security firm Kaspersky Lab researched options available in these markets, they found people offering to arrange a five-minute DDoS attack for as little as $5, with an all-day attack costing around $400. Kaspersky calculated that organising a botnet of around 1,000 computers would cost about $7 per hour. Sellers charge an average of $25 for attacks of this length, generating a healthy profit margin.[17] The year of the WannaCry attack, the dark net market for ransomware was estimated to be worth millions of dollars, with some vendors making six-figure salaries (tax-free, of course).[18]

Despite the popularity of malware with criminal groups, it's suspected that some of the most advanced examples originally evolved from government projects. When WannaCry infected susceptible computers, it did so by exploiting a so-called 'zero-day' loophole, which is when software has a vulnerability that isn't publicly known. The loophole behind WannaCry was allegedly identified by the US National Security Agency as a way of gathering intelligence, before somehow finding its way into other hands.[19] Tech companies can be willing to pay a lot to close these loopholes. In 2019, Apple offered a bounty of up to $2 million for anyone who could hack into the new iPhone operating system.[20]

During a malware outbreak, zero-day loopholes can boost transmission by increasing the susceptibility of target machines. In 2010, the 'Stuxnet' worm was discovered to have infected Iran's Natanz nuclear facility. According to later reports, this meant it would have been able to damage the vital centrifuges. To successfully spread through the Iranian systems, the worm had exploited twenty zero-day loopholes, which was almost unheard of at the time. Given the sophistication of the attack, many in the media pointed to the US and Israeli military as potential creators of the worm. Even so, the initial infection may have been the result of something far simpler: it's been suggested that the worm got into the system via a double agent with an infected USB stick.[21]

Computer networks are only as strong as their weakest links. A few years before the Stuxnet attack, hackers successfully accessed a highly fortified US government system in Afghanistan. According to journalist Fred Kaplan, Russian intelligence had supplied infected USB sticks to several shopping kiosks near the NATO headquarters in Kabul. Eventually an American soldier had bought one and used it with a secure computer.[22] It's not only humans who pose a security risk. In 2017, a US casino was surprised to discover its data had been flowing to a hacker's computer in Finland. But the real shock was the source of the leak. Rather than targeting the well-protected main server, the attacker had got in through the casino's internet-connected fish tank.[23]

HISTORICALLY, HACKERS have been most interested in accessing or disrupting computer systems. But as technology increasingly becomes internet-connected, there is growing interest in using computer systems to control other devices. This can include highly personal technology. While that casino fish tank was being targeted in Nevada, Alex Lomas and his colleagues at British security firm Pen Test Partners were wondering whether it was possible to hack into Bluetooth-enabled sex toys. It didn't take them long to discover that some of these devices were highly vulnerable to attack. Using only a few lines of code, they could in theory hack a toy and set it vibrating at its maximum setting. And because devices allow only one connection at a time, the owner would have no way of turning it off.[24]

Of course, Bluetooth devices have a limited range, so could hackers really do this in reality? According to Lomas, it's certainly possible. He once checked for nearby Bluetooth devices while walking down a street in Berlin. Looking at the list on his phone, he was surprised to see a familiar ID: it was one of the sex toys that his team had shown could be hacked. Someone was presumably carrying it with them, unaware a hacker could easily switch it on.

It's not just Bluetooth toys that are susceptible. Lomas' team found other devices were vulnerable too, including a brand of sex toy with a WiFi-enabled camera. If people hadn't changed the default password, it would be fairly easy to hack into the toy and access the video stream. Lomas has pointed out that the team has never tried to connect to a device outside their lab. Nor did they do the research to shame people who might use these toys. Quite the opposite: by raising the issue, they wanted to ensure that people could do what they wanted without fear of being hacked, and in doing so pressure the industry to improve standards.

It's not just sex toys that are at risk. Lomas has found that the Bluetooth trick also worked on his father's hearing aids. And some targets are even larger: computer scientists at Brown University discovered that it was possible to gain access to research robots, due to a loophole in a popular robotics operating system. In early 2018, the team managed to take control of a machine at the University of Washington (with the owners' permission). They also found threats closer to home. Two of their own robots – an industrial helper and a drone – were accessible to outsiders. 'Neither was intentionally made available on the public Internet,' they noted, 'and both have the potential to cause physical harm if used inappropriately.' Although the researchers focused on university-based robots, they warned that similar problems could affect machines elsewhere. 'As robots move out of the lab and into industrial and home settings, the number of units that could be subverted is bound to increase manifold.'[25]

The internet of things is creating new connections across different aspects of our lives. But in many cases, we may not realise exactly where these connections lead. This hidden network became apparent at lunchtime on 28 February 2017, when several people with internet-connected homes noticed that they couldn't turn on their lights. Or turn off their ovens. Or get into their garages.

The glitch was soon traced to Amazon Web Services (AWS), the company's cloud computing subsidiary. When a person hits the switch to turn on a smart light bulb, it will typically notify a cloud-based server – such as AWS – potentially located thousands of miles away. This server will then send a signal back to the bulb to turn it on. That February lunchtime, however, some of the AWS servers had briefly gone offline. With the server down, a large number of household devices had stopped responding.[26]

AWS has generally been very reliable – the company promises working servers over 99.99 per cent of the time – and if anything this reliability has boosted the popularity of such cloud computing services. In fact, they've become so popular that almost three-quarters of Amazon's recent profits have come from AWS alone.[27] However, widespread use of cloud computing, combined with the potential impact of a server failure, has led to suggestions that AWS might be 'too big to fail'.[28] If large amounts of the web rely on a single company, small problems at the source could be greatly amplified. Related concerns surfaced in 2018, when Facebook announced that millions of its users had been affected by a security breach. Because many people use their Facebook account to sign in to other websites, such attacks may spread further than users initially realise.[29]

This isn't the first time we've met this combination of hidden links and highly connected hubs. These are the same network quirks that made the pre-2008 financial system vulnerable, allowing seemingly local events to have an international impact. In online networks, however, these effects can be even more extreme. And this can lead to some rather unusual outbreaks.

NOT LONG AFTER THE MILLENNIUM BUG came the 'love bug'. In early May 2000, people around the world received e-mails with a subject line that read 'ILOVEYOU'. The message carried a computer worm, which was disguised as a text file containing a love letter. When opened, the worm corrupted files on that person's computer and e-mailed itself to everyone in their address book. It spread widely, crashing the e-mail system of several organisations, including the UK parliament. Eventually IT departments rolled out countermeasures, which protected computers against the worm. But then something odd happened. Rather than disappear, the worm persisted. Even a year later, it was still one of the most active bits of malware on the internet.[30]

Computer scientist Steve White had noticed the same thing happening with other computer worms and viruses. In 1998, he'd pointed out that such bugs would often linger online. 'Now here's the mystery,' White wrote.[31] 'Our evidence on virus incidents indicates that, at any given time, few of the world's systems are infected.' Although viruses persisted for a long time in the face of control measures, suggesting they were highly contagious, they generally infected relatively few computers, which implied they weren't that good at spreading.

What was causing this apparent paradox? A couple of months after the love bug attack, Alessandro Vespignani and fellow physicist Romualdo Pastor-Satorras came across White's paper. Computer viruses didn't seem to behave like biological epidemics, so the pair wondered if the structure of the network might have something to do with it. The previous year, a study had shown that there was a lot of variation in popularity on the world wide web: most websites had very few links, while some had a vast number.[32]

We've already seen that for STIs, the reproduction number of an infection will be larger when there is a lot of variation in how many sexual partners people have. An infection that would fade away if everyone behaved identically can persist if some people have a lot more partners than others. Vespignani and Pastor-Satorras realised that something even more extreme can happen with computer networks.[33] Because there is huge variability in the number of links, even seemingly weak infections can survive. The reason is that in this kind of network, a computer is never more than a few steps from a highly connected hub, which can spread the infection widely in a superspreading event. It's an exaggerated form of the problem that banks faced in 2008, with a few major hubs able to drive the entire outbreak.

When outbreaks are driven by superspreading events, it makes the transmission process extremely fragile. Unless an infection hits a major hub, it probably won't go very far. Yet superspreading can also make an outbreak more unpredictable. Although most outbreaks won't take off, those that do can stutter along for a surprisingly long time. This explains why a handful of computer viruses and worms have continued to spread, despite not being that transmissible at an individual level. The same is true of many trends on social media. If you've ever seen a strange meme spreading and wondered how it could have persisted for so long, it probably has more to do with the network itself rather than the quality of the content.[34] Thanks to their structure, online networks are giving infections an advantage that they don't have in other areas of life.

ON 22 MARCH 2017, web developers around the world noticed that their apps weren't working properly. From Facebook to Spotify, companies using the JavaScript programming language found themselves unable to work parts of their software. User interfaces were broken, visuals wouldn't load, updates wouldn't install.

The problem? Eleven lines of computer code – which many people didn't even know existed – had gone missing. The code in question had been written by Azer Koçulu, a developer based in Oakland, California. Those eleven lines formed a JavaScript program called 'left-pad'. The program itself wasn't particularly complicated; it just added some extra characters at the start of a segment of text. It was the sort of thing most coders could have created from scratch in a few minutes.[35]

Yet most coders don't create everything from scratch. To save time, they use tools that others have developed and shared. Many of them do this by searching an online resource called 'npm', which collects together handy bits of code like left-pad. In some cases, people incorporate these existing tools into new programs, which they subsequently share. Some of these programs then feed into other new programs, creating a chain of dependency with each one supporting the next. Whenever someone installs or updates a program, they will also need to load everything in the dependency chain, otherwise they'll get an error message. Left-pad lay deep within one of these chains. In the month before it disappeared, the code had been downloaded over two million times.

On that day in March, Koçulu had pulled his code from npm after a disagreement over a trademark. Npm had asked him to rename one of his software packages after another company complained; Koçulu protested and eventually responded by removing all of his code. That included left-pad, which meant that any chains of programs that relied on Koçulu's tool were suddenly broken. And because some of the chains were so long, many developers hadn't realised they were so reliant on those eleven lines of code.

Koçulu's work is just one example of computer code that has spread much further than we might think. Soon after the left-pad incident, software developer David Haney noted that another tool on npm – which consisted of a single line of code – had become an essential part of seventy-two other programs. He listed several other pieces of software that were highly dependent on simple snippets of code. 'I can't help but be amazed by the fact that developers are taking on dependencies for single line functions that they should be able to write with their eyes closed,' he wrote.[36] Borrowed pieces of code can often spread further than people realise. When researchers at Cornell University analysed articles written with LaTeX, a popular scientific writing software, they found that academics would often repurpose each other's code. Some files had spread through networks of collaborators for more than twenty years.[37]

As code spreads, it can also pick up changes. After those three students posted the Mirai code online at the end of September

2016, dozens of different variants emerged, each with subtly different features. It was only a matter of time before someone altered the code to launch a major attack. In early October, a few weeks before the Dyn incident, security company RSA noticed a remarkable claim on a dark net marketplace: a group of hackers was offering a way to flood a target with 125 gigabytes of activity per second. For $7,500, someone could buy access to a 100,000-strong botnet, which was apparently based on some adapted Mirai code.[38] However, it wasn't the first time the Mirai code had changed. In the weeks before they published the code, Mirai's creators made over twenty alterations, apparently in an attempt to increase the contagiousness of their botnet. These included features that made the worm harder to detect, as well as tweaks to fight off other malware that was competing for the same susceptible machines. Once out in the wild, Mirai would continue to change for years to come; new variants were still appearing in 2019.[39]

When Fred Cohen first wrote about computer viruses in 1984, he pointed out that malware might evolve over time, becoming harder to detect. Rather than settling down to a well-balanced equilibrium, the ecosystem of computer viruses and anti-virus software would continuously shift around. 'As evolution takes place, balances tend to change, with the eventual result being unclear in all but the simplest circumstances,' he noted.[40] 'This has very strong analogies to biological theories of evolution, and might relate well to genetic theories of diseases.'

A common way of protecting against malware is to have anti-virus software look for known threats. Typically, this involves searching for familiar segments of code; once a threat is recognised, it can be neutralised.[41] Human immune systems can do something very similar when we get infected or vaccinated. Immune cells will often learn the shape of the specific pathogen we've been exposed to; if we get infected again, these cells can respond quickly and neutralise the threat. However, evolution can sometimes hinder this process, with pathogens that once looked familiar changing their appearance to evade detection.

One of the most prominent – and frustrating – examples of this process is influenza evolution. Biologist Peter Medawar once called the flu virus 'a piece of nucleic acid surrounded by bad news'.[42] There are two particular types of bad news on the surface of the virus: a pair of proteins known as haemaggluti-nin and neuraminidase, or HA and NA for short. HA allows the virus to latch onto host cells; NA helps with the release of new virus particles from infected cells. The proteins can take several different forms, and the different flu types – like $H1N1$, $H3N2$, $H5N1$ and so on – are named accordingly.

Winter flu epidemics are mostly caused by $H1N1$ and $H3N2$. These viruses gradually evolve as they circulate, causing the shape of those proteins to change. This means our immune system no longer recognises the mutated virus as a threat. We have annual flu epidemics – and annual flu vaccination cam-paigns – because our bodies are in essence playing a game of evolutionary cat-and-mouse with the infection.

Evolution can also help artificial infections persist. In recent years, malware has started to alter itself automatically to make identification harder. During 2014, for example, the 'Beebone' botnet infected thousands of machines worldwide. The worm behind the bots changed its appearance several times a day, resulting in millions of unique variants as it spread. Even if anti-virus software learned what the current versions of code looked like, the worm would soon shuffle itself around, distorting any known patterns. Beebone was finally taken offline in 2015, when police targeted the part of the system that wasn't evolving: the fixed domain names used to co-ordinate the botnet. This proved far more effective than trying to identify the shapeshift-ing worms.[43] Similarly, biologists are hoping to develop more effective flu vaccines by targeting the parts of the virus that don't change.[44]

Given the need to evade detection, malware will continue to evolve, while authorities attempt to keep up. The routes of transmission will also keep changing. As well as finding new targets – like household devices – infections are increasingly spreading through clickbait and tailored attacks on social media.[45] By sending customised messages to specific users, hackers can boost the chances they'll click on a link and inadvertently let malware in. However, evolution isn't just helping infections spread effectively from computer-to-computer or person-to-person. It's also revealing a new way to tackle contagion.

Tracking outbreaks

THE AFFAIR WOULD END with a murder attempt. For over ten years, Richard Schmidt, a gastroenterologist in Lafayette, Louisiana, had been having a relationship with Janice Trahan, a nurse fifteen years his junior. She'd divorced her husband after the affair started, but despite his promises, Schmidt had not left his wife and three children. Trahan had tried to break off the affair before, but this time it would be for good.

She would later testify that a couple of weeks afterwards, on 4 August 1994, Schmidt had come to her home while she was asleep. Schmidt told her he was there to give her a shot of vitamin B12. He'd previously given her vitamin injections to boost her energy levels, but that night she told him she didn't want one. Before she could stop him, he'd stuck a needle in her arm. None of the previous injections had hurt, but this time the pain spread right through the limb. At which point, Schmidt said he had to leave to go to the hospital.

The pain continued overnight, and in the weeks that followed, she became ill with flu-like symptoms. She made several trips to the hospital, but test after test came back negative. One doctor had suspected HIV, but didn't test for it. He later said that his colleague – one Dr Schmidt – had told him that Trahan had already tested negative for the infection. Her illness continued, and eventually another doctor ordered a new set of tests. In January 1995, Trahan finally received the correct diagnosis: she was HIV positive.

Back in August, Trahan had told a colleague she'd suspected that the 'shot in the dark' wasn't B12. There was no doubt that HIV was a recent infection: she'd given blood several times and her most recent donation – made in April 1994 – had tested negative for HIV. According to a local HIV specialist, the progression of her symptoms was consistent with an early August date of infection. When police searched Schmidt's offices, they found evidence that blood had been drawn from an HIV patient on 4 August – just hours before he'd allegedly injected Trahan – and the procedure hadn't been recorded in the usual way. However, Schmidt denied visiting her and giving her the injection.[1]

Perhaps the virus itself could provide a clue about what had happened? At the time, it was already common to use DNA testing to match suspects to crime scenes. However, the task was trickier in this case. Viruses like HIV evolve relatively quickly, so the virus found in Trahan's blood wouldn't necessarily be the same as the one in the blood that infected her. Faced with a charge of attempted second-degree murder, Schmidt argued that the HIV virus that infected Trahan was too different to the original patient's virus; it just wasn't plausible that this had been the source of her infection. Given all the other evidence pointing to Schmidt, the prosecution disagreed. They just needed a way to show it.

ON 20 JUNE 1837, THE BRITISH CROWN passed down the royal family tree, from William IV to Victoria. Meanwhile, a short walk away in Soho, a young biologist was also thinking about family trees, albeit on a much grander scale. Back in England after his five-year voyage on HMS *Beagle*, Charles Darwin would end up outlining his theories in a new leather-bound notebook. To

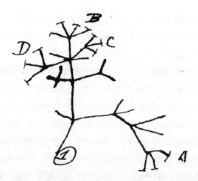

Darwin's original tree of life sketch. Species A is a distant relative
of B, C, and D, which are more closely related. In the diagram,
all the species evolved from a single starting point, labelled (1)

help clarify his thinking, he sketched out a simplified diagram
of a 'tree of life'. The idea was that the branches indicated the
evolutionary relationships between different species. Just like
a family tree, Darwin suggested that closely related organisms
would be closer to each other, while distinct species would be
much further away. Tracing each of the branches would lead to
a shared root: a single common ancestor.

Darwin started by drawing evolutionary trees based on
things like physical traits. On his *Beagle* voyage, he catego-
rised bird species by features such as beak shape, tail length,
and plumage.[2] This field of research would eventually become
known as 'phylogenetics', after the Ancient Greek words for
'species' (phylo) and 'origin' (genesis).

Although early evolutionary analysis focused on the appearance of different species, the rise of genetic sequencing has
made it possible to compare organisms in much more detail. If
we have two genomes, we can see how related they are based on
the amount of overlap in the lists of letters that make up their
sequences. The more overlap there is, the fewer mutations are
required to get from one sequence to the other. It's a bit like

waiting for tiles to appear in a game of Scrabble. Going from a sequence 'AACG' to 'AACC', for example, is easier than getting from 'AACG' to 'TTGG'. And like Scrabble, we can estimate how long the evolutionary process has been running based on how much the letters have changed from their original sequence.

Using this idea – and plenty of computational power – it's possible to arrange sequences into a phylogenetic tree, tracing out their historical evolution. We can also estimate when important evolutionary changes may have happened. This is useful if we want to know how an infection may have spread. For example, after SARS sparked a major outbreak in 2003, scientists identified the virus in palm civets, a small mongoose-like animal. Maybe the disease had been routinely circulating in civets before spilling over into the human population?

Analysis of different SARS viruses suggested otherwise. Human and civet viruses were closely related, indicating that both were relatively new hosts for the virus. SARS had potentially jumped from civets into humans a few months before the outbreak started. In contrast, the virus had been circulating in bats for much longer, making its way into civets sometime around 1998. Based on the evolutionary history of the different viruses, civets were probably just a brief stepping stone for SARS as it made its way into humans.[3]

During Richard Schmidt's trial, the prosecution used similar phylogenetic evidence to show that it was plausible that Trahan's infection had come from the HIV patient who'd visited Schmidt. Evolutionary biologist David Hillis and his colleagues compared the viruses isolated from the pair with other viruses found in HIV patients in Lafayette. In his testimony, Hillis said the viruses found in Schmidt's patient and Trahan were 'the most closely related sequences in the analysis, and as closely related to sequences isolated that two individuals could be'. Although it wasn't conclusive proof that Trahan's infection had come from Schmidt's patient, it undermined the defence's claim

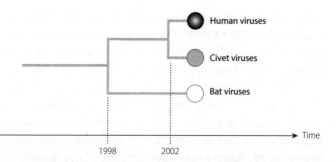

Simplified phylogenetic tree for SARS viruses in different
host species. Dashed lines show estimated times when
viruses diverged from one another, finding their way
into a new group of hosts. (Data: Hon et al., 2008)

that the cases were unrelated. Eventually, Schmidt was found
guilty and sentenced to fifty years in prison. As for Trahan, she
remarried and continued to live with HIV, celebrating her twentieth wedding anniversary in 2016.[4]

Schmidt's trial was the first time that phylogenetic analysis
had been used in a US criminal case. Since then, the methods
have appeared in other cases around the world. Following a
surge in cases of hepatitis C in Valencia, Spain, police investigators linked many of the patients to an anaesthetist named
Juan Maeso. Phylogenetic analysis confirmed he was the likely
source of the outbreak, and in 2007 he was convicted of infecting hundreds of patients by reusing syringes.[5] Genetic data has
also helped prove innocence. Shortly after the Maeso case, a
group of medics were released from a prison in Libya. They'd
been held for eight years after accusations that they'd deliberately infected children with HIV. The group were freed in part
because of phylogenetic analysis, which showed that many of
the infections had occurred years before the team had arrived
in the country.[6]

As well as pointing to the likely source of an outbreak,

phylogenetic methods can reveal when a disease arrived in a particular location. Suppose we are investigating a virus like HIV, which evolves relatively quickly. If the HIV viruses circulating in an area are relatively similar, it suggests they haven't had long to evolve, so the outbreak is probably quite recent. In contrast, if there is a lot of diversity among current viruses, it means that there has been a lot of time for evolution, which suggests the original virus was introduced a while ago. These methods are now commonly used in public health. Recall how in earlier chapters, we looked at the arrival of Zika into Latin America and HIV into North America. In both cases, teams used genetic data to estimate the timing of the virus's introduction. Researchers have also applied these same ideas to other infections, from pandemic influenza to hospital superbugs like MRSA.[7]

With access to genetic data, we can also work out whether an outbreak started with a single case or multiple introductions. When our team analysed Zika viruses isolated in Fiji during 2015 and 2016, we found two distinct groups of viruses in the phylogenetic tree. Based on the rate of evolution, one group of viruses had arrived into the capital Suva in 2013–14, spreading at low levels for the subsequent year or two, while a separate outbreak had later started in the west of the country.[8] I didn't realise it at the time, but some of the mosquitoes I swatted away during my 2015 visit had probably been infected with Zika.

Another benefit of phylogenetic analysis is that we can track transmission in the final stages of an outbreak. In March 2016, a new cluster of Ebola cases appeared in Guinea, three months after WHO had declared the West Africa epidemic over. Perhaps the virus had been spreading undetected in humans all along? When epidemiologist Boubacar Diallo and his collaborators sequenced viruses from the new cluster of cases, they hit upon an alternative explanation. The new viruses were closely related to an Ebola virus found in the semen of a local man who'd recovered from the disease back in 2014. The virus had persisted in his body for almost a year-and-a-half, before spreading to a sexual partner and sparking a new outbreak.[9]

Sequence data is becoming an important part of outbreak analysis, but the idea of evolving viruses can sometimes lead to alarmist coverage. During the Ebola and Zika epidemics, several media reports played up the fact that the viruses were evolving.[10] But this isn't necessarily as bad as it sounds: all viruses evolve, in the sense that their genetic sequence gradually changes over time. Occasionally this evolution will lead to a difference we care about – like the flu virus changing its appearance – but often it will just happen in the background without having a noticeable effect on an outbreak.

The rate of evolution can affect our ability to analyse outbreaks, though. Phylogenetic analysis is more effective when looking at pathogens that evolve fairly quickly, like HIV and flu. This is because the genetic sequence will change as pathogens spread from one person to another, making it possible to estimate the likely path of infection. In contrast, viruses like measles evolve slowly, which means there won't be much variation from one person to another.[11] As a result, working out how the cases are related is a bit like trying to piece together a human family tree in a country where everyone has the same surname.

As well as biological limitations to phylogenetic methods, there are also practical ones. In the early stages of the West Africa Ebola epidemic, Pardis Sabeti, a geneticist with the Broad Institute in Boston, analysed sequence data from ninety-nine viruses from Sierra Leone. Phylogenetic trees showed that the infection had spread from Guinea to Sierra Leone in May 2014, possibly after a funeral. Given the seriousness of the outbreak, Sabeti and her colleagues quickly added the new genetic sequences to a public database. This initial burst of research was then followed by a period of relative silence. Although several other teams had been collecting virus samples, nobody else released any new genetic sequences between 2 August and 9 November 2014. During this same period, there were over 10,000 Ebola cases reported in West Africa, with the epidemic reaching its peak in October.[12]

There are a couple of possible reasons for the delay in releasing sequences. The cynical explanation is that new data are valuable academic currency. Research papers using genetic sequences to study outbreaks are likely to get published in coveted scientific journals, which creates an incentive for researchers to sit on potentially important data. However, based on my interactions with researchers during this period, I'd like to think it was mostly a matter of obliviousness rather than malicious intentions. Scientific culture just wasn't adapted for outbreak timelines. Researchers are used to developing protocols, performing thorough analysis, writing up their methods, submitting the results to be peer-reviewed by fellow scientists. This process can take months – if not years – and has historically slowed the release of new data.

Such delays are a problem across science and medicine. When Jeremy Farrar took over as director of the Wellcome Trust in March 2014, he told *The Guardian* that clinical research often took too long, something that became apparent in the following months as the Ebola outbreak grew. 'The systems we have got in place are not fit for purpose when the situation is moving quickly,' Farrar said. 'We have nothing that enables us to respond in real time.'[13]

This culture is gradually changing. In mid-2018, what would become another major Ebola outbreak began in the Democratic Republic of the Congo. This time, researchers were quick to release new sequence data. Teams also launched a clinical trial of four experimental treatments. By August 2019, they'd shown that a prompt infusion of anti-Ebola immune cells could increase someone's chances of survival to over 90 per cent, up from a historical average of around 30 per cent. Meanwhile, outbreak scientists are increasingly posting draft papers on websites like bioRxiv and medRxiv, which aim to make new research accessible before it undergoes peer-review.[14]

During her time working in Sierra Leone, Sabeti discovered that the word for Kenema, the city where they were based, meant 'clear like a river, translucent and open to the public gaze'.[15] This openness was reflected in her team's work, with those ninety-nine sequences shared early in the outbreak. The attitude has also taken hold among the wider community of outbreak researchers. One of the best examples is the Nextstrain project, pioneered by computational biologists Trevor Bedford and Richard Neher. This online platform automatically collates genetic sequences to show how different viruses are related and where they might have come from. Although Bedford and Neher initially focused on flu, the platform now tracks everything from Zika to tuberculosis.[16] Nextstrain has proved to be a powerful idea, not just because it brings together and visualises all the available sequences, but because it's separate from the slow and competitive process of publishing scientific papers.

In early 2020, Nextstrain would provide a crucial real-time indication that COVID-19 was a bigger problem in the US than the initial data suggested.[17] On 27 February, a teenager had been diagnosed with the disease in Mill Creek, near Seattle where Bedford is based. Prior to that point, only a handful of cases had been reported in the state. Within twenty-four hours, the coronavirus from Mill Creek had been sequenced and the results posted online, where they were picked up by the Nextstrain system. The initial phylogenetic tree suggested COVID-19 had been spreading undetected for a while. Bedford posted a thread on Twitter describing the analysis, which would be widely shared.[18] 'I believe we're facing an already substantial outbreak in Washington State that was not detected until now due to narrow case definition requiring direct travel to China,' he concluded. It didn't take long for the theory to be verified: over the following days, cases and deaths started climbing sharply in the state.

As it becomes easier to sequence pathogens, phylogenetic methods will continue to improve our understanding of disease outbreaks. They will help us discover when infections first sparked, how outbreaks grew, and what parts of a transmission process we might have missed. The methods also illustrate a wider trend in outbreak analysis: the ability to combine new data sources to get at information that has traditionally been hard to come by. With phylogenetics, we can uncover the spread of outbreaks by linking patient information with the genetic data of the viruses that infected them. These kinds of 'data linkage' approaches are becoming a powerful way of understanding how things mutate and spread in a population. But they aren't always being used in the ways we might expect.

GOLDILOCKS WAS A DISHONEST, foul-mouthed old woman who burgled a trio of well-meaning bears. At least, she was when poet Robert Southey first published the story in 1837. After swearing her way through three bowls of porridge and breaking a chair, the woman heard the bears come home and made her escape through a window. Southey didn't give her a name or golden hair; those details would come decades later, as the villainous woman evolved into a troublesome child and finally the Goldilocks most of us know today.[19]

The tale of the bears has been around for a long time. A few years before Southey published his story, a woman named Eleanor Mure had written a homemade book for her nephew. This time the bears caught the old woman at the end of the story. Angry at the damage, the bears set her on fire, tried to drown her, and then impaled her on the steeple of St Paul's Cathedral. In an earlier folk story, three bears saw off a mischievous fox.

According to Jamie Tehrani, an anthropologist at Durham University, we can think of culture as information that mutates as it gets transmitted from person-to-person and generation-to-generation. If we want to understand the spread and evolution of culture, folk stories are therefore useful because they are the product of their society. 'By definition, folktales don't have a single authoritative version,' said Tehrani. 'They are stories that belong to everybody in the community. They have this organic quality.'[20]

Tehrani's work on folktales started with 'Little Red Riding Hood'. If you live in Western Europe, you're probably familiar with the tale as told by the Brothers Grimm in the nineteenth century: a girl visits her grandmother's house, only to be met by a wolf in disguise. However, this isn't the only version of the story. There are several other folk tales out there that bear similarities to 'Little Red Riding Hood'. In Eastern Europe and the Middle East, people tell the story of 'The Wolf and the Kids': a disguised wolf tricks a group of baby goats into letting him into their house. In East Asia, there is the tale of 'The Tiger Grandmother', in which a group of children encounter a tiger that pretends to be their elderly relative.

The tale has spread across the world, but it's difficult to tell in which direction. A common theory among historians is that the East Asian version was the original, with the European and Middle Eastern stories coming later. But did 'Little Red Riding Hood' and 'The Wolf and the Kids' really evolve from 'The Tiger Grandmother'? Folktales have historically been spoken rather than written down, which means historical records are shallow and patchy. It's often not clear exactly when and where a particular story originated.

This is where phylogenetic approaches can come in useful. To investigate the evolution of 'Little Red Riding Hood' and its variants, Tehrani gathered together almost sixty different versions of the story, spanning multiple continents. In place of a genetic sequence, he summarised each story based on a set of seventy-two plot features, such as the type of lead character, the trick used to deceive them, and how the story ended. He then estimated how these features evolved, resulting in a phylogenetic tree that mapped the relationship between the stories.[21] His analysis would produce an unexpected conclusion: based on the phylogenetic tree, it seemed that 'The Wolf and the Kids' and 'Little Red Riding Hood' had come first. Contrary to common belief, 'The Tiger Grandmother' was apparently a blend of existing tales, rather than being the original version from which others evolved.

Evolutionary thinking has a long history in the study of language and culture. Decades before Darwin drew his tree of life, linguist William Jones had been interested in how languages emerge, a field known as 'philology'. In 1786, Jones noted the similarities between Greek, Sanskrit, and Latin: 'no philologer could examine them all three, without believing them to have sprung from some common source, which, perhaps, no longer exists.'[22] In evolutionary terms, he was suggesting that these languages had evolved from a single common ancestor. Jones's ideas would later influence many other scholars, including the Brothers Grimm, who were keen linguists. As well as collecting together different variants of folktales, they tried to study how the use of language had changed over time.[23]

Modern phylogenetic methods make it possible to analyse the evolution of such stories in much more detail. After studying 'Little Red Riding Hood', Jamie Tehrani worked with Sara Graça da Silva at the University of Lisbon to examine a much wider range of stories, tracing the evolution of 275 folktales in total. The pair found that some tales have a long history; stories such as 'Rumplestiltskin' and 'Beauty and the Beast' may have originally emerged over 4,000 years ago. This would mean they are as old as the Indo-European languages through which they spread. Although many folktales eventually travelled widely, da Silva and Tehrani also found traces of local rivalry in story-telling. 'Spatial proximity appears to have had a negative effect on the tales' distributions,' they noted, 'suggesting that societies were more likely to reject than adopt these stories from their neighbours.'[24]

Folktales are often tied to a country's identity, even if their origins are not. When the Brothers Grimm compiled their collection of traditional 'German' stories, they noticed that there were similarities with tales in many other cultures, from Indian to Arabic. Phylogenetic analysis confirms just how much story borrowing there has been. 'There's not a great deal that's special about any one country's oral tradition,' Tehrani said. 'In fact, they're highly globalised.'

Why did humans start telling stories in the first place? One explanation is that tales help us preserve useful information. There's evidence that storytelling is a highly valued skill in hunter-gatherer societies, leading to suggestions that stories took hold in the early stages of human history because good storytellers were more desirable as mates.[25] There are two competing theories about what sort of story-based information we have evolved to value. Some researchers suggest that stories relating to survival are most important: deep down, we want information about where food and dangers are. This would explain why tales that evoke reactions like disgust are memorable; we don't want to poison ourselves. Others have argued that because social interactions dominate human life, socially relevant information is most useful. This would imply that we preferentially remember details about relationships and actions that break social norms.[26]

To test these two theories, Tehrani and his colleagues once ran an experiment looking at the spread of urban legends. Their study mimicked the children's game of 'broken telephone': tales were passed from one person to another, then to another, with the final version showing how much was remembered.

They found that stories containing survival or social information were more memorable than neutral stories, with the social
stories outperforming the survival ones.

Other factors can also boost the success of stories. Earlier
broken telephone experiments found that tales tend to become
shorter and simpler as they spread: people remember the gist
but forget the details. Surprises can help a tale as well. There's
evidence that tales are catchier if they include counter-intuitive
ideas. However, there is a balance to be struck. Stories need
some surprising features, but not too many. Successful folk
tales generally have a lot of familiar elements, combined with a
couple of absurd twists. Take Goldilocks, the story of a girl who
explores the family home of a mother, father, and baby. The
twist, of course, being that it's a family of bears. This narrative
trick also explains the attraction of conspiracy theories, which
take real-life events and add an unexpected slant.[27]

Then there's the structure of a story. Goldilocks' popularity
might not be down to her, but rather the three bears. They turn
the story into a sequence of memorable triplets: the bowls of
porridge are too hot, too cold, just right; the beds are too soft,
too hard, just right. This rhetorical trick is known as the 'rule of
three' and crops up regularly in politics, from the speeches of
Abraham Lincoln to Barack Obama.[28] Why are lists of three so
powerful? It might have something to do with the mathematical
importance of triplets: in general, we need at least three items
in a sequence to establish (or break) a pattern.[29]

Patterns can also help with the spread of individual words.
As language evolves, new words often have to compete to dis-
place already popular ones. In such situations, we might expect
people to prefer words that follow consistent rules. For example,
past tense verbs often end in '...ed', so it makes sense that the
historical word 'smelt' has made way for 'smelled', while 'wove'
is gradually becoming 'weaved'.[30]

Yet some words have evolved in the other direction. In the

1830s, people would have 'lighted' a candle; nowadays we'd talk of having lit one. Why did these irregular words outcompete popular ones? A group of biologists and linguists at the University of Pennsylvania reckon that rhyming might have had something to do with it. They noticed that in the mid-twentieth century, Americans started saying 'dove' instead of 'dived' as the past tense of 'to dive'. Around the same time, newly popular cars were causing people to adopt words like 'drive' and 'drove'. Similarly, people started using 'lit' and 'quit' instead of 'lighted' and 'quitted' during the period that 'split' became a popular way of saying you were going to leave.

There are two main ways that new words and stories can spread through a population. Either they pass down from generation-to-generation, perhaps picking up some variations along the way; this is known as 'vertical transmission'. Alternatively, tales may blend across communities in the same generation, in a process of 'horizontal transmission'. Da Silva and Tehrani have found that both types of transmission have influenced the spread of folktales, but for the majority of stories, the vertical route was more important. In other areas of life though, horizontal transmission can dominate. Creators of computer programs often reuse existing lines of code, perhaps because there's a useful feature they need to include, or because they want to save time. In evolutionary terms, this means that computer code can 'time travel', with bits of old programs or languages suddenly popping up in new ones.[31]

If sections of stories or computer code mix together within a single generation, it becomes difficult to draw a neat evolutionary tree. If a parent tells their child a traditional family story, then the child incorporates parts of their friends' family stories, the new tale essentially fuses all these different branches of stories together. The same problem is well known to biologists. Take the 2009 'swine flu' pandemic. The outbreak started when genes from four viruses – a bird flu virus, a human flu virus and two different swine flu strains – jumbled together inside an infected pig in Mexico, creating a new hybrid virus that then spread among humans.[32] One gene was closely related to other human flu viruses; another was similar to circulating bird flu strains; others were like swine viruses. And yet, taken as a whole, this new flu virus wasn't really like anything else. Changes like these show the limitations of a simple tree metaphor. Although Darwin's tree of life captures many features of evolution, the reality – with genes potentially passing within as well as between generations – is more like a bizarre, unkempt hedge.[33]

The processes of horizontal and vertical transmission can make a big difference to how traits spread through a population. In the waters of Shark Bay, just off the coast of Western Australia, a handful of bottlenose dolphins have started using tools to forage for food. Marine biologists first noticed the behaviour in 1984; dolphins were breaking off bits of marine sponge and wearing them as a protective mask while they rummaged for fish in the seabed. But not all dolphins in Shark Bay would go on to use 'sponging'. Only around one in ten have picked up the technique.[34] Why hasn't the behaviour spread further? Twenty years after biologists first observed sponging, a group of researchers used genetic data to show that the tactic was almost entirely the result of vertical transmission. Dolphins are famously social, but it seems that after one initial dolphin came up with the innovation, it only spread through their family line. Individuals who weren't related to them kept on foraging sponge-free. In effect, this family of dolphins had created their own unique tradition.

According to ecologist Lucy Aplin, both vertical and horizontal transmission of culture can occur in the animal world. 'It really depends on the species, and also on the behaviour being learned.' She points out that the type of transmission can affect how widely new information spreads. 'You might imagine in, say, dolphins, where most of the learning happens vertically, you end up with family-specific behaviours and it's quite hard for behaviours to spread more widely through the population.' In contrast, horizontal transmission can result in much faster adoption of innovations. Such transmission is common in species of birds like great tits. 'Much of their social learning occurs horizontally,' Aplin said, 'with information gained by observing unrelated individuals in the winter-flocking period, rather than transmitted from parent to offspring.'[35]

For some animals, the difference between transmission types could prove crucial to survival. As humans alter the natural environment more and more, species that can efficiently transmit innovations will be better placed to adjust to the changes. 'Evidence is increasingly showing that some species can show a high degree of behavioural flexibility in the face of changing environments,' Aplin said. 'As a result, they appear to be successful at coping with human-modified habitats and human-induced change.'

Efficient transmission is also helping organisms resist human change at the microscopic level. Several types of bacteria have picked up mutations that make them resistant to antibiotics. As well as spreading vertically when bacteria reproduce, these genetic mutations often pass horizontally within the same generation. Just as software developers might copy and paste code between files, bacteria can pick up snippets of genetic material from each other. In recent years, researchers have discovered that this horizontal transmission is contributing to the emergence of superbugs such as MRSA, as well as drug-resistant STIs.[36] As bacteria evolve, many common infections may eventually become untreatable. In 2018, for example, a man in the UK was diagnosed with so-called 'super-gonorrhea', which was resistant to all standard antibiotics. He'd picked up the infection in Asia, but the following year two more cases appeared in the UK, this time with links to Europe.[37] If researchers are to successfully track and prevent such infections, they will need all the data they can get.

THANKS TO THE AVAILABILITY of new information sources like genetic sequences, we are increasingly able to unravel how different diseases and traits spread through populations. Indeed, one of the biggest changes to human healthcare in the twenty-first century will be the ability to rapidly and cheaply sequence and analyse genomes. As well as uncovering outbreaks, researchers will be able to study how human genes influence conditions ranging from Alzheimer's to cancer.[38] Genetics has social applications too. Because our genomes can reveal characteristics like ancestry, genetic testing kits have become popular gifts for people interested in their family history.

Yet the availability of such data can have unintended effects on privacy. Because we share so many genetic characteristics with our relatives, it's possible to learn things about people who haven't been tested. In 2013, for example, *The Times* reported that Prince William had Indian ancestry, after testing two distant cousins on his mother's side. Genetics researchers soon criticised the story, because it had revealed personal information about the prince without his consent.[39] In some cases ancestry revelations can have devastating consequences: there have been several reports of families thrown into disarray after discovering hidden adoptions or infidelity in a Christmas ancestry test.[40]

We've already seen how data about our online behaviour is gathered and shared so that companies can target adverts. Marketers don't just measure how many people clicked on an ad; they know what kind of person they are, where they came from, and what they did next. By combining these datasets, they can piece together how one thing influences another. The same approach is common when analysing human genetic data. Rather than look at genetic sequences in isolation, scientists will compare them with information like ethnic background or medical history. The aim is to uncover the patterns that link the different datasets. If researchers know what these look like, they can predict things like ethnicity or disease risk from the underlying genetic code. This is why genetic testing companies like 23andMe have attracted so many investors. They aren't just collecting customers' genetic data; they are gathering information about who these people are, which makes it possible to gain much deeper health insights.[41]

It's not just for-profit companies that are building such datasets. Between 2006 and 2010, half a million people volunteered for the UK Biobank project, which aims to study patterns in genetics and health over the coming decades. As the dataset grows and expands, it will be accessible to teams around the globe, creating a valuable scientific resource. Since 2017, thousands of researchers have signed up to access the data, with projects investigating diseases, injuries, nutrition, fitness, and mental health.[42]

There are huge benefits to sharing health information with researchers. But if datasets are going to be accessible to multiple groups, we need to think about how to protect people's privacy. One way to reduce this risk is to remove information that could be used to identify participants. For example, when researchers get access to medical datasets, personal information like name and address will often have been removed. Even without such data, though, it may still be possible to identify people. When Latanya Sweeney was a graduate student at MIT in the mid-1990s, she suspected that if you knew a US citizen's age, gender, and ZIP code, in many cases you could narrow it down to a single person. At the time, several medical databases included these three pieces of information. Combine them with an electoral register and Sweeney reckoned you could probably work out whose medical records you were looking at.[43]

So that's what she did. 'To test my hypothesis, I needed to look up someone in the data,' she later recalled.[44] The state of Massachusetts had recently made 'anonymised' hospital records freely available to researchers. Although Governor William Weld had claimed the records still protected patients' privacy,

Sweeney's analysis suggested otherwise. She paid $20 to access voter records for Cambridge, where Weld lived, then cross-referenced his age, gender, and ZIP code against the hospital dataset. She soon found his medical records, then mailed him a copy. The experiment – and the publicity it generated – would eventually lead to major changes in how health information is stored and shared in the US.[45]

As data spread from one computer to another, so do the resulting insights into people's lives. It's just not medical or genetic information we need to be careful with; even seemingly innocuous datasets can hold surprisingly personal details. In March 2014, a self-described 'data junkie' named Chris Whong used the Freedom of Information Act to request details of every yellow taxi ride in New York City during the previous year. When the New York City Taxi and Limousine Commission released the dataset, it included the time and location of the pick up and drop off, the fare, and how much each passenger tipped.[46] There were over 173 million trips in total. Rather than give the real licence plates, each taxi was identified by a string of apparently random digits. But it turned out the journeys were anything but anonymous. Three months after the dataset was released, computer scientist Vijay Pandurangan showed how to decipher the taxi codes, converting the scrambled digits back into the original licence plates. Then graduate student Anthony Tockar published a blog post explaining what else could be discovered. He'd found that with a few simple tricks, it was possible to extract a lot of sensitive information from the files.[47]

First, he showed how a person might stalk celebrities. After hours spent trawling through a search of images for 'celebrities in taxis in Manhattan in 2013', Tockar found several pictures with a licence plate in view. Cross-referencing these with celebrity blogs and magazines, he worked out what the start point or destination was, and matched this against the supposedly anonymous taxi dataset. He could also see how much celebrities had

– or hadn't – tipped. 'Now while this information is relatively benign, particularly a year down the line,' Tockar wrote, 'I have revealed information that was *not previously in the public domain.*'

Tockar acknowledged that most people might not be too worried about such analysis, so he decided to dig a little further. He turned his attention to a strip club in the Hell's Kitchen neighbourhood, searching for taxi pick-ups in the early hours. He soon identified a frequent customer and tracked the person's journey back to their home address. It didn't take long to find them online and – after a quick search on social media – Tockar knew what the man looked like, how much his house was worth, and what his relationship status was. Tockar chose not to publish any of this information, but it wouldn't have taken much effort for someone else to come to the same conclusions. 'The potential consequences of this analysis cannot be overstated,' Tockar noted.

With high-resolution GPS data, it can be extremely easy to identify people.[48] Our GPS tracks can easily reveal where we live, what route we take to work, what appointments we have, and who we meet. As with the New York Taxi data, it doesn't take much to spot how such information could be a potential treasure trove for stalkers, burglars, or blackmailers. In a 2014 survey, 85 per cent of US domestic violence shelters said they were protecting people from abusers who'd stalked them via GPS.[49] Consumer GPS data can even put military operations at risk. During 2017, army staff wearing commercial fitness trackers inadvertently leaked the exact layout of bases when they uploaded their running and cycling routes.[50]

Despite these risks, the availability of movement data is also bringing valuable scientific insights, whether it's allowing researchers to estimate where viruses might spread next, helping emergency teams support displaced populations after natural disasters, or showing planners how to improve city transport networks.[51] With high-resolution GPS data, it's even becoming possible to analyse interactions between specific groups of people. For example, studies have used mobile phone data to track social segregation, political groupings and inequality in countries ranging from the United States to China.[52]

If that last sentence made you feel slightly uncomfortable, you wouldn't be alone. As the availability of digital data increases, concerns about privacy are growing too. Issues like inequality are a major social challenge – and undoubtedly worthy of research – but there is intense debate about how far such research should delve into the details of our incomes, politics or social lives. When it comes to understanding human behaviour, we often have a decision to make: what is an acceptable price for knowledge?

Whenever my collaborators and I have worked on projects involving movement data, privacy has been hugely important to us. On the one hand, we want to collect the most useful data we possibly can, especially if it could help to protect communities against outbreaks. On the other, we need to protect the private lives of the individuals in those communities, even if this means limiting the information we collect or publish. For diseases like flu or measles, we face a particular challenge, because children – who are at high risk of infection – are also a vulnerable age group to be putting under surveillance.[53] There are plenty of studies that could tell us useful, interesting things about social behaviour, but would be difficult to justify given the potential infringement on privacy.

In the rare instances where we do go out and collect high-resolution GPS data, our study participants will have given consent and know that only our team will have access to their exact location. But not everyone has the same attitude to privacy. Imagine if your phone had been leaking GPS data continuously, without your knowledge, to companies you've never heard of. This is more likely than you might think. In recent years, a little-known network of GPS data brokers has emerged. These companies have been buying movement data from hundreds of apps that people have given GPS access, then selling this on to marketers, researchers and other groups.[54] Many users may have long forgotten they installed these apps – be it for fitness, weather forecasts or gaming – let alone agreed to constant tracking. In 2019, US journalist Joseph Cox reported that he'd paid a bounty hunter to track a phone using second-hand location data.[55] It had cost $300.

As location data becomes easier to access, it is also inspiring new types of crimes. Scammers have long used 'phishing' messages to trick customers into giving sensitive information. Now they are developing 'spear phishing' attacks, which incorporate user-specific data. In 2016, several residents of Pennsylvania, USA received e-mails asking them to pay a fine for a recent speeding offence. The e-mails correctly listed the speed and location of the person's car. But they weren't real. Police suspected that scammers had obtained leaked GPS data from an app, then used this to identify people who'd been travelling too fast on local roads.[56]

Although movement datasets are proving remarkably powerful, they do still have some limitations. Even with very detailed movement information, there is one type of interaction that is near impossible to measure. It's an event that is brief, often invisible, and particularly elusive in the early stages of outbreak. It's also one that has sparked some of the most notorious incidents in medical history.

THE DOCTOR CHECKED INTO ROOM 911 of Hong Kong's Metropole Hotel at the end of a tiring week. Despite feeling unwell, he'd made the three-hour bus trip across from Southern China for his nephew's wedding that weekend. He'd come down with a flu-like illness a few days earlier and hadn't managed to shake it off. However, it was about to get much worse. Twenty-four hours later, he'd be in an intensive care unit. Within ten days, he would be dead.[57]

It was 21 February 2003, and the doctor was the first case of SARS in Hong Kong. Eventually, there would be sixteen other SARS cases linked to the Metropole: people who'd stayed in rooms opposite the doctor, beside him, or along the corridor. As the disease spread, there was an urgent need to understand the new virus causing it. Scientists didn't even know basic information like the delay from infection to appearance of symptoms (i.e. the incubation period). With cases appearing across Southeast Asia, statistician Christl Donnelly and her colleagues at Imperial College London and in Hong Kong set out to estimate this crucial information.[58]

The problem with working out an incubation period is that we rarely see the actual moment of infection. We just see people showing up with symptoms later on. If we want to estimate the average incubation period, we therefore need to find people who could only have been infected during a specific period of time. For example, a businessman staying at the Metropole had overlapped with the Chinese doctor for a single day. He fell ill with SARS six days later, so this delay must have been the incubation period for his infection. Donnelly and her colleagues tried to gather together other examples like this, but there weren't that many. Of the 1,400 SARS cases that had been reported in Hong Kong by the end of April, only 57 people had a clearly defined exposure to the virus. Put together, these examples suggested that SARS had an average incubation period of about 6.4 days. The same method has since been used to estimate the incubation period for other new infections, including pandemic flu in 2009, Ebola in 2014 and COVID-19 in 2020.[59]

Of course, there is another way to work out an incubation period: deliberately give someone the infection and see what happens. One of the most infamous examples of this approach occurred in New York City during the 1950s and 1960s. The

Willowbrook State School, located on Staten Island, was home to over 6,000 children with intellectual disabilities. Overcrowded and filthy, the school had frequent outbreaks of hepatitis, which had led paediatrician Saul Krugman to set up a project to study the infection.[60] Working with collaborators Robert McCollum and Joan Giles, the research involved deliberately infecting children with hepatitis to understand how the infection developed and spread. As well as measuring the incubation period, the team discovered they were actually dealing with two different types of hepatitis virus. One type, which we now call hepatitis A, spread from person-to-person, whereas hepatitis B was blood-borne.

The research brought controversy as well as discoveries. In the early 1970s, criticism of the work grew, and the experiments were eventually halted. The study team argued that the project had been ethically sound: it had approval from several medical ethics boards, they'd obtained consent from childrens' parents, and the poor conditions in the school meant that many of the children would have got the disease at some point anyway. Critics responded that, among other things, the consent forms had brushed over the details of what was involved and Krugman overstated the chances children would get infected naturally. 'They were the most unethical medical experiments ever performed on children in the United States,' claimed vaccine pioneer Maurice Hillman.[61]

This raises the question of what to do with such knowledge once it's been obtained. Research papers from the Willowbrook study have been cited hundred of times, but not everyone agreed they should be acknowledged in this way. 'Every new reference to the work of Krugman and Giles adds to its apparent ethical respectability, and in my view such references should stop, or at least be heavily qualified,' wrote physician Stephen Goldby in a letter to The Lancet in 1971.[62]

There are many other examples of medical knowledge that

has uncomfortable origins. In early nineteenth-century Britain, the growing number of medical schools created a massive demand for cadavers for use in anatomy classes. Faced with a limited legal supply, the criminal market stepped in; bodies were increasingly snatched from graveyards and sold to lecturers.[63] Yet it is experiments on the living that have proved the most shocking. During the Second World War, Nazi doctors deliberately infected patients at Auschwitz with diseases including typhus and cholera, to measure things like the incubation period.[64] After the war, the medical community created the Nuremberg Code, outlining a set of principles for ethical studies. Even so, the controversies would continue. Much of our understanding of typhoid comes from studies involving US prisoners in the 1950s and 1960s.[65] Then, of course, there was Willowbrook, which transformed our knowledge of hepatitis.

Despite the sometimes horrific history of human experiments, studies involving deliberate infections are on the rise.[66] Around the world, volunteers are signing up for research involving malaria, influenza, dengue fever, and others. In 2019, there were dozens of such studies underway. Although some pathogens are simply too dangerous – Ebola is clearly out of the question – there are situations in which the social and scientific benefits of an infection experiment can outweigh a small risk to participants. Modern infection experiments have much stricter ethical guidelines, particularly when giving participants information and asking for their consent, but they must still strike this balance between benefit and risk. It's a balancing act that is becoming increasingly prominent in other areas of life as well.

8

A spot of trouble

GRENVILLE CLARK HAD JUST ABOUT SETTLED into his position as conference chair when someone handed him a folded note.[1] A lawyer by training, Clark had organised the conference to discuss the future of the newly formed United Nations and what it would mean for world peace. Sixty delegates had already arrived at the Princeton University venue, but there was one more person who wanted to join. The note in Clark's hands came from Albert Einstein, who was based at the adjacent Institute for Advanced Studies.

It was January 1946, and many in the physics community were haunted by their role in the recent atomic bombings of Hiroshima and Nagasaki.[2] Although Einstein was a long-time pacifist – and had opposed the bombings – his letter to President Roosevelt in 1939, warning of the potential for a Nazi atom bomb, had triggered the US nuclear programme.[3] During the Princeton conference, one attendee asked Einstein about humanity's inability to manage new technology.[4] 'Why is it that when the mind of man has stretched so far as to discover the structure of the atom we have been unable to devise the political means to keep the atom from destroying us?' 'That is simple, my friend,' replied Einstein. 'It is because politics is more difficult than physics.'

Nuclear physics is one of the most prominent examples of a 'dual-use technology'.[5] The research has brought huge scientific and social benefits, but it has also found extremely harmful uses. In the preceding chapters, we've met several other examples of technology that can have both a positive and negative use. Social media can connect us to old friends and useful new ideas. Yet it can also enable the spread of misinformation and other harmful content. Analysis of crime outbreaks can identify people who may be at risk, making it possible to interrupt transmission; it can also feed into biased policing algorithms that may over-target minority groups. Large-scale GPS data is revealing how to respond effectively to catastrophes, how to improve transport systems, and how new diseases might spread.[6] But it also risks leaking personal information without our knowledge, endangering our privacy and even our safety.

In March 2018, the *Observer* newspaper reported that Cambridge Analytica had secretly gathered data from tens of millions of Facebook users, with the aim of building psychological profiles of US and British voters.[7] Although the effectiveness of such profiling has been disputed by statisticians,[8] the scandal eroded public trust in technology firms. According to software engineer – and ex-physicist – Yonatan Zunger, the story was a modern retelling of the ethical debates that had already occurred in fields like nuclear physics or medicine.[9] 'The field of computer science, unlike other sciences, has not yet faced serious negative consequences for the work its practitioners do,' he wrote at the time. As new technology appears, we mustn't forget the lessons that researchers in other fields have already learned the hard way.

When 'big data' became a popular buzzword in the early twenty-first century, the potential for multiple uses was a source of optimism. The hope was that data collected for one purpose could help tackle questions in other areas of life. A flagship example of this was Google Flu Trends (GFT).[10] By analysing the search patterns of millions of users, researchers suggested it would be possible to measure flu activity in real-time, rather than waiting a week or two for official US disease tallies to be published.[11] The initial version of GFT was announced in early 2009, with promising results. However, it didn't take long for criticisms to emerge.

The GFT project had three main limitations. First, the predictions didn't always work that well. GFT had reproduced the seasonal winter flu peaks in the US between 2003 and 2008, but when the pandemic took off unexpectedly in spring 2009, GFT massively underestimated its size.[12] 'The initial version of GFT was part flu detector, part winter detector,' as one group of academics put it.[13]

The second problem was that it wasn't clear how the predictions were actually made. GFT was essentially an opaque machine; search data went in one end and predictions came out the other. Google didn't make the raw data or methods available to the wider research community, so it wasn't possible for others to pick apart the analysis and work out why the algorithm performed well in some situations but badly in others.

Then there's the final – and perhaps biggest – issue with GFT: it didn't seem that ambitious. We get flu epidemics each winter because the virus evolves, making current vaccines less effective. Similarly, the main reason governments are so worried about a future pandemic flu virus is that we won't have an effective vaccine against the new strain. In the event of a pandemic, it would take six months to develop one,[14] by which time the virus will have spread widely. To predict the shape of flu outbreaks, we need a better understanding of how viruses evolve, how people interact, and how populations build immunity.[15] Faced with this hugely challenging situation, GFT merely aimed to report flu activity a week or so earlier than it would have been otherwise. It was an interesting idea in terms of data analysis, but not a revolutionary one when it comes to tackling outbreaks.

This is a common pitfall when researchers or companies talk about applying large datasets to wider aspects of life. The tendency is to assume that, because there is so much data, there must be other important questions it can answer. In effect, it becomes a solution in search of a problem.

IN LATE 2016, epidemiologist Caroline Buckee attended a tech fundraising event, pitching her work to Silicon Valley insiders. Buckee has a lot of experience of using technology to study outbreaks. In recent years, she has worked on several studies using GPS data to investigate malaria transmission. But she is also aware that such technology has its limitations. During the fundraising event, she became frustrated by the prevailing attitude that with enough money and coders, companies could solve the world's health problems. 'In a world where technology moguls are becoming major funders of research, we must not fall for the seductive idea that young, tech-savvy college grads can single-handedly fix public health on their computers,' she wrote afterwards.[16]

Many tech approaches are neither feasible nor sustainable. Buckee has pointed to many failed attempts at tech pilot studies or apps that hoped to 'disrupt' traditional methods. Then there's the need to evaluate how well health measures actually work, rather than just assuming good ideas will emerge naturally like successful start-ups. 'Pandemic preparedness requires a long-term engagement with politically complex, multidimensional problems – not disruption,' as she put it.

Technology can still play a major role in modern outbreak analysis. Researchers routinely use mathematical models to help design control measures, smartphones to collect patient data, and pathogen sequences to track the spread of infection.[17] However, the biggest challenges are often practical rather than computational. Being able to gather and analyse data is one thing; spotting an outbreak and having the resources to do something about it is quite another. When Ebola caused its first major epidemic in 2014, transmission was centred on Sierra Leone, Liberia and Guinea, three countries that ranked among the world's poorest. A second major epidemic would begin in 2018, when Ebola hit a conflict zone in the northeastern part of the Democratic Republic of the Congo; by July 2019, with 2,500 cases and rising, WHO would declare it a Public Health Emergency of International Concern (PHEIC).[18] The global imbalance in health capacity even shows up in scientific terminology. The 2009 pandemic flu virus emerged in Mexico, but its official designation is 'A/California/7/2009(H1N1)', because that's where a lab first identified the new virus.[19]

These logistical challenges mean that research can struggle to keep up with new outbreaks. During 2015 and 2016, Zika spread widely, spurring researchers to plan large-scale clinical studies and vaccine trials.[20] But as soon as many of these studies were ready to start, the cases stopped. This is a common frustration in outbreak research; by the time the infections end, fundamental questions about contagion can remain unanswered. That's why building long-term research capacity is essential. Although our research team has managed to generate a lot of data on the Zika outbreak in Fiji, we were only able to do this because we already happened to be there investigating dengue. Similarly, some of the best data on Zika have come from a long-running Nicaraguan dengue study led by Eva Harris at the University of California, Berkeley.[21]

Researchers have also lagged behind outbreaks in other fields. Many studies of misinformation during the 2016 US election weren't published until 2018 or 2019. Other research projects looking at election interference have struggled to get off the ground at all, while some are now impossible because social media companies – whether inadvertently or deliberately – have deleted the necessary data.[22] At the same time, fragmented and unreliable data sources are hindering research into banking crises, gun violence and opioid use.[23]

Getting data is only part of the problem, though. Even the best outbreak data will have quirks and caveats, which can hinder analysis. In her work tracking radiation and cancer, Alice Stewart noted that epidemiologists rarely have the luxury of a perfect dataset. 'You're not looking for a spot of trouble against a spotless backdrop,' she said,[24] 'you're looking for a spot of trouble in a very messy situation.' The same issue crops up in many fields, whether trying to estimate the spread of obesity in friendship data, uncover patterns of drug use in the opioid epidemic, or trace the effects of information across different social media platforms. Our lives are messy and complicated, and so are the datasets they produce.

If we want a better grasp of contagion, we need to account for its dynamic nature. That means tailoring our studies to different outbreaks, moving quickly to ensure our results are as useful as possible, and finding new ways to thread strands of information together. For example, disease researchers are now combining data on cases, human behaviour, population immunity, and pathogen evolution to investigate elusive outbreaks. Taken individually, each dataset has its own flaws, but together they can reveal a more complete picture of contagion. Describing such approaches, Caroline Buckee has quoted Virginia Woolf, who once said that 'truth is only to be had by laying together many varieties of error.'[25]

As well as improving the methods we use, we should also focus on the questions that really matter. Take social contagion. Considering the amount of data now available, our understanding of how ideas spread is still remarkably limited. One reason is that the outcomes we care about aren't necessarily the ones that technology companies prioritise. Ultimately, they want users to interact with their products in a way that brings in advertising revenue. This is reflected in the way we talk about online contagion. We tend to focus on the metrics designed by social media companies ('How do I get more likes? How do I get this post to go viral?') rather than outcomes that will actually make us healthier, happier, or more successful.

With modern computational tools, there is potential to get unprecedented insights into social behaviour, if we target the right questions. The irony, of course, is that the questions we care about are also the ones that are likely to lead to controversy. Recall that study looking at the spread of emotions on Facebook, in which researchers altered people's News Feeds to show happier or sadder posts. Despite criticism of how this research was designed and carried out, the team was asking an important question: how does the content we see on social media affect our emotional state?

Emotions and personality are, by their very definition, emotive and personal topics. In 2013, psychologist Michal Kosinski and his colleagues published a study suggesting that it was possible to predict personality traits – such as extroversion and intelligence – from the Facebook pages that people liked.[26] Cambridge Analytica would later use a similar idea to profile voters, triggering widespread criticism.[27] When Kosinski and his team first published their method, they were aware that it could have uncomfortable alternative uses. In their original paper, they even anticipated a possible backlash against technology firms. The researchers speculated that as people became more aware of what could be extracted from their data, some might turn away from digital technology entirely.

If users are uncomfortable with exactly how their data is being used, researchers and companies have two options. One is to simply avoid telling them. Faced with concerns about privacy, many tech companies have downplayed the extent of data collection and analysis, fearing negative press coverage and uproar from users. Meanwhile, data brokers (who most of us have never heard of) have been making money selling data (which we weren't aware they had) to external researchers (who we didn't know were analysing it). In these cases, the assumption seems to have been that if you tell people what you're doing with their data, they won't let you do it. Thanks to new privacy laws like Europe's General Data Protection Regulation (GDPR) and California's Consumer Privacy Act, some of these activities are becoming harder. But if research teams continue to brush over the ethics of their analysis, there will be further scandals and lapses in trust. Users will become more reluctant to share data, even for worthwhile studies, and researchers will shy away from the effort and controversy of analysing it.[28] As a result, our understanding of behaviour – and the social and health benefits that can come from such insights – will stagnate.

The alternative option is to increase transparency. Instead of analysing people's lives without their knowledge, let them weigh up the benefits and risks. Involve them in the debates; think in terms of permission rather than forgiveness. If social benefits are the aim, make the research a social effort. When the NHS announced their 'Care.data' scheme in 2013, the hope was that better data sharing could lead to better health research. Three years later, the scheme was cancelled after the public – and doctors – lost confidence in how the data were being used. In theory, Care.data could have been enormously beneficial, but patients didn't seem to know about the scheme, or didn't trust it.[29]

Perhaps nobody would agree to data-intensive research if they knew what was really involved? In my experience, that's not necessarily true. Over the past decade, my collaborators and I have run several 'citizen science' projects combining contagion research with wider discussions about outbreaks, data, and ethics. We've studied what networks of interactions look like, how social behaviour changes over time, and what this means for infection patterns.[30] Our most ambitious project was a massive data collection effort we ran in collaboration with the BBC during 2017/18.[31] We asked the public to download a smartphone app that tracked their movements to the nearest 1km over a day, and also asked them to tally up their social interactions. Once the study was completed, this dataset would help form a freely available resource for researchers. To our surprise, tens of thousands of people volunteered, despite the project having no immediate benefit to them. Although just one study, it shows that large-scale data analysis can still be carried out in a transparent and socially beneficial way.

In March 2018, the BBC aired a program called *Contagion!*, showcasing the initial dataset we'd gathered. It wasn't the only story about large-scale data collection in the media that week; a few days earlier, the Cambridge Analytica scandal had broken. Whereas we had asked people to volunteer their data to help researchers understand disease outbreaks, Cambridge Analytica had allegedly been harvesting vast quantities of Facebook data – without users' knowledge – to help politicians try and influence voters.[32] Here were two studies of behaviour, two massive datasets, and two very different outcomes. Several commentators picked up on the contrast, including journalist Hugo Rifkind in his TV review for *The Times*. 'In a week when we've agreed that data and internet surveillance – boo, hiss – are ruining the world, *Contagion* was a welcome reminder that it can sort of save it a bit too.'[33]

Almost two years later, on 29 February 2020, the first report of local COVID-19 transmission in the UK would emerge from the town Haslemere in Surrey.[34] By coincidence, this was the same town we had chosen to spark a fictional simulated pandemic in the *Contagion* program. It meant we knew what to expect next: if there was transmission in Haslemere, outbreaks would probably be taking off in London within a week or two. Sure enough, within ten days there were reports that a member of Parliament had been infected, an early sign of what was a rapidly growing epidemic. During the following months, my colleagues and I would use the BBC dataset to investigate several aspects of transmission in the UK, from the potential effectiveness of new contact tracing strategies to the changing social behaviour of a country under lockdown.[35]

Despite the importance of knowing how people interact if we want to stop an epidemic, the shadow of Cambridge Analytica would loom over discussions of data and privacy throughout COVID-19 response. It would influence which forms of disease surveillance were deemed acceptable, from contact tracing apps to requirements to 'check-in' when visiting bars and restaurants. Meanwhile, the online manipulation playbook, honed in the 2010s on political content and campaigns against vaccines, would find a natural home in the resulting Venn diagram of a highly polarised pandemic. Across the world, responses would be shaped – and often stifled – by the misinformation and disinformation that spread alongside the virus itself. Politics, it seems, is also more difficult than epidemiology.

IN THE TIME IT'S TAKEN you to read this book, around three hundred people will have died of malaria. There will have been over five hundred deaths from HIV/AIDS, and about eighty from measles, most of them children. Melioidosis, a bacterial infection that you may well have never heard of, will have killed more than sixty people.[36]

Infectious diseases still cause vast damage worldwide. As well as known threats, we face the ever-present risk of a new pandemic – such as COVID-19 – and the rising emergence of drug-resistant infections. However, as our knowledge of contagion has improved, infectious diseases have on the whole declined. The global death rate for such diseases has halved in the past two decades.[37]

As infectious diseases wane, attention is gradually shifting to other threats, many of which can also be contagious. In 1950, tuberculosis was the leading cause of death for a British man in his thirties. Since the 1980s, it has been suicide.[38] In recent years, young adults in Chicago have been most likely to die from homicide.[39] Then there are the wider social burdens of contagion. When I analysed neknomination back in 2014, online transmission seemed like a tangential issue, almost a curiosity. Three years later, it was dominating front pages, with concerns about the spread of false information – and the role of social media – leading to multiple government investigations.[40]

As our awareness of contagion increases, many of the ideas honed in the study of infectious diseases are now translating to other types of outbreaks. After the 2008 financial crisis, central banks latched onto the idea that the structure of a network could amplify contagion, a theory pioneered by STI researchers in the 1980s and 1990s. Recent efforts to treat violence as an infection – rather than simply a result of 'bad people' – echo the rejection of diseases caused by 'bad air' in the 1880s and 1890s. Concepts like the reproduction number are helping researchers quantify the spread of innovations and online content, while methods used to study pathogen sequences are revealing the transmission and evolution of culture. Along the way, we're finding new ways to speed up beneficial ideas and slow down harmful ones. Just as Ronald Ross hoped in 1916, a modern 'theory of happenings' is now helping us analyse everything from diseases and social behaviour to politics and economics.

In many cases, this has meant overturning popular notions of how outbreaks work. Like the idea that we need to remove every last mosquito to control malaria, or vaccinate every person to prevent epidemics. Or the assumption that banking systems are naturally stable and online content is highly contagious. It has also meant hunting for new explanations: why cases of Guillain-Barré Syndrome were appearing on Pacific Islands, why computer viruses persist for so long, why most ideas struggle to spread as easily as diseases.

In outbreak analysis, the most significant moments aren't the ones where we're right. It's those moments when we realise we've been wrong. When something doesn't look quite right: a pattern catches our eye, an exception breaks what we thought was the rule. Whether we want an innovation to take off or an infection to decline, these are the moments we need to reach as early as possible. The moments that allow us to unravel chains of transmission, searching for weak links, missing links, and unusual links. The moments that let us look back, to work out how outbreaks really happened in the past. Then look forward, to change how they happen in future.

Notes

Introduction

1. WHO, 'Coronavirus disease 2019 (COVID-19) Situation Report –
   24', 13 February 2020.
2. Kucharski A. J. et al., 'Early dynamics of transmission and control
   of COVID-19: a mathematical modelling study', Lancet Inf Dis,
   2020.
3. Gale J., 'Coronavirus May Infect Up to 500,000 in Wuhan Before It
   Peaks', Bloomberg, 8 February 2020.
4. Cyranoski D., 'When will the coronavirus outbreak peak?' Nature,
   18 February 2020.
5. The Novel Coronavirus Pneumonia Emergency Response
   Epidemiology Team. 'The Epidemiological Characteristics of
   an Outbreak of 2019 Novel Coronavirus Diseases (COVID-19) –
   China, 2020', China CDC Weekly, 2020.
6. Russell T. W. et al., 'Estimating the infection and case fatality ratio
   for COVID-19 using age-adjusted data from the outbreak on the
   Diamond Princess cruise ship', MedRxiv, 9 March 2020. (Later
   published in Eurosurveillance.)
7. Linde P., 'Uncontrolled spread of coronavirus in Spain forces a
   change of scenario', El Pais, 10 March 2020.
8. Hayes A, Nair A., 'Coronavirus: Sixth person dies in UK as
   confirmed cases rise to 382', Sky News, 10 Mar 2020; Jit M et al.
   'Estimating number of cases and spread of coronavirus disease

(COVID-19) using critical care admissions, United Kingdom, February to March 2020', *Eurosurveillance*, 2020.

9. Leclerc Q. et al., 'What settings have been linked to SARS-CoV-2 transmission clusters?' *Wellcome Open Research*, 2020.

10. Kucharski A. J. et al., 'Early dynamics of transmission and control of COVID-19: a mathematical modelling study', Lancet Inf Dis, 2020.

11. Lancet Global Health Lab, 17 March 2020. Available from: https://www.lshtm.ac.uk/newsevents/events/virtual-event-what-best-way-stop-pandemic.

12. Glenny M., 'Coronavirus is boomtime for computer hackers: how to keep your tech secure when working from home', *Sunday Times*, 24 May 2020.

13. Background on 1918 pandemic: Barry J.M., 'The site of origin of the 1918 influenza pandemic and its public health implications' *Journal of Translational Medicine*, 2004; Johnson N.P.A.S. and Mueller J., 'Updating the Accounts: Global Mortality of the 1918–1920 "Spanish" Influenza Pandemic' *Bulletin of the History of Medicine*, 2002; World War One casualty and death tables. PBS, Oct 2016. https://www.uwosh.edu/faculty_staff/henson/188/WWI_Casualties%20and%20Deaths%20%20PBS.html. Note that there have recently been other theories about the source of the 1918 flu pandemic, with some arguing that the introduction was much earlier than previously thought e.g. Branswell H., 'A shot-in-the-dark email leads to a century-old family treasure – and hope of cracking a deadly flu's secret', *STAT News*, 2018.

14. Examples of quote in media: Gerstel J., 'Uncertainty over H1N1 warranted, experts say' *Toronto Star*, 9 October 2009; Osterholm M.T., 'Making sense of the H1N1 pandemic: What's going on?' Center for Infectious Disease Research and Policy, 2009.

15. Eames K.T.D. et al., 'Measured Dynamic Social Contact Patterns Explain the Spread of H1N1v Influenza', *PLOS Computational Biology*, 2012; Health Protection Agency, 'Epidemiological report of pandemic (H1N1) 2009 in the UK', 2010.

16. Other groups reached similar conclusions, e.g. WHO Ebola Response Team, 'Ebola Virus Disease in West Africa – The First

9 Months of the Epidemic and Forward Projections', *The New England Journal of Medicine (NEJM)*, 2014.

17. 'Ransomware cyber-attack: Who has been hardest hit?', BBC News Online, 15 May 2017; 'What you need to know about the WannaCry Ransomware', Symantec Blogs, 23 October 2017. Exploit attempts increased from 2000 to 80000 in 7 hours, implying doubling time = $7/\log_2(80000/2000) = 1.32$ hours.

18. Media Metrics #6: The Video Revolution. The Progress & Freedom Foundation Blog, 2 March 2008. http://blog.pff.org/archives/2008/03/print/005037.html. Adoption went from 2.2% of homes in 1981 to 18% homes in 1985, implying doubling time = $365 \times 4/\log_2(0.18/0.02) = 481$ days.

19. Etymologia: influenza. *Emerging Infectious Diseases* 12(1):179, 2006.

1. A theory of happenings

1. Dumas A., *The Count of Monte Cristo* (1844–46), Chapter 117.

2. Kucharski A.J. et al., 'Using paired serology and surveillance data to quantify dengue transmission and control during a large outbreak in Fiji', *eLIFE*, 2018.

3. Pastula D.M. et al., 'Investigation of a Guillain-Barré syndrome cluster in the Republic of Fiji', *Journal of the Neurological Sciences*, 2017; Musso D. et al., 'Rapid spread of emerging Zika virus in the Pacific area', *Clinical Microbiology and Infection*, 2014; Sejvar J.J. et al., 'Population incidence of Guillain-Barré syndrome: a systematic review and meta-analysis', *Neuroepidemiology*, 2011.

4. Willison H.J. et al., 'Guillain-Barré syndrome', *The Lancet*, 2016.

5. Kron J., 'In a Remote Ugandan Lab, Encounters With the Zika Virus and Mosquitoes Decades Ago', *New York Times*, 5 April 2016.

6. Amorim M. and Melo A.N., 'Revisiting head circumference of Brazilian newborns in public and private maternity hospitals', *Arquivos de Neuro-Psiquiatria*, 2017.

7. World Health Organization, 'WHO statement on the first meeting of the International Health Regulations (2005) (IHR 2005) Emergency Committee on Zika virus and observed increase in neurological disorders and neonatal malformations', 2016.

8. Rasmussen S.A. et al., 'Zika Virus and Birth Defects – Reviewing the Evidence for Causality', *NEJM*, 2016.

9. Rodrigues L.C., 'Microcephaly and Zika virus infection', *The Lancet*, 2016.

10. Unless otherwise stated, background information is from: Ross R., *The Prevention of Malaria* (New York, 1910); Ross R., Memoirs, *With a Full Account of the Great Malaria Problem and its Solution* (London, 1923).

11. Barnes J., *The Beginnings Of The Cinema In England, 1894–1901: Volume 1: 1894–1896* (University of Exeter Press, 2015).

12. Joy D.A. et al., 'Early origin and recent expansion of Plasmodium falciparum', *Science*, 2003.

13. Mason-Bahr P., 'The Jubilee of Sir Patrick Manson: A Tribute to his Work on the Malaria Problem', *Postgraduate Medical Journal*, 1938.

14. To K.W.K. and Yuen K-Y., 'In memory of Patrick Manson, founding father of tropical medicine and the discovery of vector-borne infections' *Emerging Microbes and Infections*, 2012.

15. Burton R., *First Footsteps in East Africa* (London, 1856).

16. Hsu E., 'Reflections on the "discovery" of the antimalarial *qinghao*', *British Journal of Clinical Pharmacololgy*, 2006.

17. Sallares R., *Malaria and Rome: A History of Malaria in Ancient Italy* (Oxford University Press, 2002).

18. Ross claimed that the participants had been told what was involved, and that risks of the experiments were justified: 'I think myself justified in making this experiment because of the vast importance a positive result would have and because I have a specific in quinine always at hand.' (source: Ross, 1923). However, it is not clear how fully the risks were actually explained to participants; quinine is not as effective as the treatments used in modern studies of malaria (source: Achan J. et al., 'Quinine, an old anti-malarial drug in a modern world: role in the treatment of malaria' *Malaria Journal*, 2011.) We will look at the ethics of human experiments in more detail in Chapter 7.

19. Bhattacharya S. et al., 'Ronald Ross: Known scientist, unknown man', *Science and Culture*, 2010.

20. Chernin E., 'Sir Ronald Ross vs. Sir Patrick Manson: A Matter of Libel', *Journal of the History of Medicine and Allied Sciences*, 1988.

21. Manson-Bahr P., *History Of The School Of Tropical Medicine In London, 1899–1949*, (London, 1956).

22. Reiter P., 'From Shakespeare to Defoe: Malaria in England in the Little Ice Age', *Emerging Infectious Diseases*, 2000.

23. High R., 'The Panama Canal – the American Canal Construction', *International Construction*, October 2008.

24. Griffing S.M. et al., 'A historical perspective on malaria control in Brazil', *Memórias do Instituto Oswaldo Cruz*, 2015.

25. Jorland G. et al., *Body Counts: Medical Quantification in Historical and Sociological Perspectives* (McGill-Queen's University Press, 2005).

26. Fine P.E.M., 'John Brownlee and the Measurement of Infectiousness: An Historical Study in Epidemic Theory', *Journal of the Royal Statistical Society, Series A*, 1979.

27. Fine P.E.M., 'Ross's *a priori* Pathometry – a Perspective', *Proceedings of the Royal Society of Medicine*, 1975.

28. Ross R., 'The Mathematics of Malaria', *The British Medical Journal*, 1911.

29. Reiter P., 'From Shakespeare to Defoe: Malaria in England in the Little Ice Age', *Emerging Infectious Diseases*, 2000.

30. McKendrick background from: Gani J., 'Anderson Gray McKendrick', *StatProb: The Encyclopedia Sponsored by Statistics and Probability Societies*.

31. Letter GB 0809 Ross/106/28/60. Courtesy, Library & Archives Service, London School of Hygiene & Tropical Medicine. © Ross Family.

32. Letter GB 0809 Ross/106/28/112. Courtesy, Library & Archives Service, London School of Hygiene & Tropical Medicine. © Ross Family.

33. Heesterbeek J.A., 'A Brief History of Ro and a Recipe for its Calculation', *Acta Biotheoretica*, 2002.

34. Kermack background from: Davidson J.N., 'William Ogilvy Kermack', *Biographical Memoirs of Fellows of the Royal Society*, 1971; Coutinho S.C., 'A lost chapter in the pre-history of algebraic analysis: Whittaker on contact transformations', *Archive for History of Exact Sciences*, 2010.

35. Kermack W.O. and McKendrick A.G., 'A Contribution to the

Mathematical Theory of Epidemics', *Proceedings of the Royal Society A*, 1927.

36. Fine P.E.M., 'Herd Immunity: History, Theory, Practice', *Epidemiologic Reviews*, 1993; Farewell V. and Johnson T., 'Major Greenwood (1880–1949): a biographical and bibliographical study', *Statistics in Medicine*, 2015.

37. Dudley S.F., 'Herds and Individuals', *Public Health*, 1928.

38. Hendrix K.S. et al., 'Ethics and Childhood Vaccination Policy in the United States', *American Journal of Public Health*, 2016.

39. Fine P.E.M., 'Herd Immunity: History, Theory, Practice', *Epidemiologic Reviews*, 1993.

40. Mallet H-P. et al., 'Bilan de l'épidémie à virus Zika survenue en Polynésie française, 2013–14', *Bulletin d'information sanitaires, épidémiologiques et statistiques*, 2015.

41. Duffy M.R. et al., 'Zika Virus Outbreak on Yap Island, Federated States of Micronesia' *NEJM*, 2009.

42. Cao-Lormeau V.M. et al., 'Guillain-Barré Syndrome outbreak associated with Zika virus infection in French Polynesia: a case-control study', *The Lancet*, 2016.

43. Stoddard S.T. et al., 'House-to-house human movement drives dengue virus transmission', *PNAS*, 2012.

44. Kucharski A.J. et al., 'Transmission Dynamics of Zika Virus in Island Populations: A Modelling Analysis of the 2013–14 French Polynesia Outbreak', *PLOS Neglected Tropical Diseases*, 2016.

45. Faria N.R. et al., 'Zika virus in the Americas: Early epidemiological and genetic findings', *Science*, 2016.

46. Andronico A. et al., 'Real-Time Assessment of Health-Care Requirements During the Zika Virus Epidemic in Martinique', *American Journal of Epidemiology*, 2017.

47. Rozé B. et al., 'Guillain-Barré Syndrome Associated With Zika Virus Infection in Martinique in 2016: A Prospective Study', *Clinical Infectious Diseases*, 2017.

48. Fine P.E.M., 'Ross's *a priori* Pathometry – a Perspective', *Proceedings of the Royal Society of Medicine*, 1975.

49. Ross R., 'An Application of the Theory of Probabilities to the

Study of a priori Pathometry – Part I', *Proceedings of the Royal Society A*, 1916.

50. Clarke B., 'The challenge facing first-time buyers', *Council of Mortgage Lenders*, 2015.

51. Rogers E.M., *Diffusion of Innovations*, 3rd Edition, (New York, 1983).

52. Background from: Bass F.M., 'A new product growth for model consumer durables', *Management Science*, 1969.

53. Bass F.M. Comments on 'A New Product Growth for Model Consumer Durables', *Management Science*, 2004.

54. Ross' simple 'susceptible-infected' model can be written as: $dS/dt = -bSI$, $dI/dt = bSI$, where b is the infection rate. The peak rate of new infections occurs when dI/dt is increasing fastest, i.e. the second derivative of dI/dt is equal to zero. Using the product rule, we obtain: $I = (3 - \text{sqrt}(3))/6 = 0.21$.

55. Jackson A.C., 'Diabolical effects of rabies encephalitis', Journal of NeuroVirology, 2016.

56. Robinson A. et al., 'Plasmodium-associated changes in human odor attract mosquitoes', *PNAS*, 2018.

57. Van Kerckhove K. et al., 'The Impact of Illness on Social Networks: Implications for Transmission and Control of Influenza', *American Journal of Epidemiology*, 2013.

58. Hudson background from: O'Connor J.J. et al., 'Hilda Phoebe Hudson', JOC/EFR, 2002; Warwick A., *Masters of Theory: Cambridge and the Rise of Mathematical Physics* (University of Chicago Press, 2003).

59. Hudson H., 'Simple Proof of Euclid II. 9 and 10', *Nature*, 1891.

60. Chambers S., 'At last, a degree of honour for 900 Cambridge women', *The Independent*, 30 May 1998.

61. Ross R. and Hudson H., 'An Application of the Theory of Probabilities to the Study of *a priori* Pathometry. Part II and Part III', *Proceedings of the Royal Society A*, 1917.

62. Letter GB 0809 Ross/161/11/01. Courtesy, Library & Archives Service, London School of Hygiene & Tropical Medicine. © Ross Family; Aubin D. et al., 'The War of Guns and Mathematics: Mathematical Practices and Communities in France and Its

Western Allies around World War I', *American Mathematical Society*, 2014.

63. Ross R., 'An Application of the Theory of Probabilities to the Study of *a priori* Pathometry. Part I', *Proceedings of the Royal Society A,*, 1916.

2. Panics and pandemics

1. Mathematician Andrew Odlyzko points out that the final loss could plausibly have been even higher than £20,000. What's more, he suggests a multiple of 1,000 is reasonable for converting monetary value in 1720 to a present day amount; Newton's Professorial salary at Cambridge during this time was around £100 per year. Source: Odlyzko A., 'Newton's financial misadventures in the South Sea Bubble', *Notes and Records, The Royal Society*, 2018.

2. Background on Thorp and Simons from: Patterson S., *The Quants* (Crown Business New York, 2010). Background on LTCM from: Lowenstein R., *When Genius Failed: The Rise and Fall of Long Term Capital Management* (Random House, 2000).

3. Allen F. et al., 'The Asian Crisis and the Process of Financial Contagion', *Journal of Financial Regulation and Compliance*, 1999. Data on rise in popularity of the term 'financial contagion' from Google Ngram.

4. Background on CDOs from: MacKenzie D. et al., '"The Formula That Killed Wall Street"? The Gaussian Copula and the Cultures of Modelling', 2012.

5. 'Deutsche Bank appoints Sajid Javid Head of Global Credit Trading, Asia', *Deutsche Bank Media Release*, 11 October 2006; Roy S., 'Credit derivatives: Squeeze is over for EM CDOs', *Euromoney*, 27 July 2006; Herrmann J., 'What Thatcherite union buster Sajid Javid learned on Wall Street', *The Guardian*, 15 July 2015.

6. Derman E., 'Model Risk' *Goldman Sachs Quantitative Strategies Research Notes*, April 1996.

7. CNBC interview, 1 July 2005.

8. According to MacKenzie et al (2012): 'The crisis was caused not by "model dopes", but by creative, resourceful, well-informed and reflexive actors quite consciously exploiting the role of models in governance.' They quote several examples of people gaming the

calculations to ensure that CDOs appeared both profitable and low-risk.

9. Tavakoli J., 'Comments on SEC Proposed Rules and Oversight of NRSROs', Letter to Securities and Exchange Commission, 13 February 2007.

10. MacKenzie D. et al., '"The Formula That Killed Wall Street"? The Gaussian Copula and the Cultures of Modelling', 2012.

11. *New Directions for Understanding Systemic Risk* (National Academies Press, Washington DC, 2007).

12. Chapple S., 'Math expert finds order in disorder, including stock market', *San Diego Union-Tribune*, 28 August 2011.

13. May R., 'Epidemiology of financial networks. Presentation at LSHTM John Snow bicentenary event, April 2013. Available on YouTube.

14. For background on May's involvement see previous note.

15. 'Was tulipmania irrational?' *The Economist*, 4 October 2013.

16. Goldgar A., 'Tulip mania: the classic story of a Dutch financial bubble is mostly wrong', *The Conversation*, 12 February 2018.

17. Online Etymology Dictionary. Origin and meaning of bubble. https://www.etymonline.com/word/bubble.

18. Reproduced with authors' permission. Source: Frehen R.G.P. et al., 'New Evidence on the First Financial Bubble', *Journal of Financial Economics*, 2013.

19. Frehen R.G.P. et al., 'New Evidence on the First Financial Bubble', *Journal of Financial Economics*, 2013.

20. Odlyzko A., 'Newton's financial misadventures in the South Sea Bubble', *Notes and Records, The Royal Society*, 2018.

21. Odlyzko A., 'Collective hallucinations and inefficient markets: The British Railway Mania of the 1840s', 2010.

22. Kindleberger C.P. et al., *Manias, Panics and Crashes: A History of Financial Crises* (Palgrave Macmillan, New York, 1978).

23. Chow E.K., 'Why China Keeps Falling for Pyramid Schemes', *The Diplomat*, 5 March 2018; 'Pyramid schemes cause huge social harm in China', *The Economist*, 3 February 2018.

24. Rodrigue J-P., 'Stages of a bubble', extract from *The Geography*

of Transport Systems (Routledge, New York, 2017). https://
transportgeography.org/?page_id=9035.

25. Sornette D. et al., 'Financial bubbles: mechanisms and diagnostics', *Review of Behavioral Economics*, 2015.

26. Coffman K.G. et al., 'The size and growth rate of the internet', *First Monday*, October 1998.

27. Odlyzko A., 'Internet traffic growth: Sources and implications', 2000.

28. John Oliver on cryptocurrency: 'You're not investing, you're gambling', *The Guardian*, 12 March 2018.

29. Data from: https://www.coindesk.com/price/bitcoin. Price was $19,395 on 18 December 2017 and $3,220 on 16 December 2018.

30. Rodrigue J-P., 'Stages of a bubble', extract from *The Geography of Transport Systems* (Routledge, New York, 2017). https://
transportgeography.org/?page_id=9035.

31. Kindleberger C.P. et al., *Manias, Panics and Crashes: A History of Financial Crises* (Palgrave Macmillan, New York, 1978).

32. Odlyzko A., 'Collective hallucinations and inefficient markets: The British Railway Mania of the 1840s', 2010.

33. Sandbu M., 'Ten years on: Anatomy of the global financial meltdown', *Financial Times*, 9 August 2017.

34. Alessandri P. et al., 'Banking on the State', *Bank of England Paper*, November 2009.

35. Elliott L. and Treanor J., 'The minutes that reveal how the Bank of England handled the financial crisis', *The Guardian*, 7 January 2015.

36. Author interview with Nim Arinaminpathy, August 2017.

37. Brauer F., 'Mathematical epidemiology: Past, present, and future', *Infectious Disease Modelling*, 2017; Bartlett M.S., 'Measles Periodicity and Community Size', *Journal of the Royal Statistical Society. Series A*, 1957.

38. Heesterbeek J.A., 'A Brief History of R0 and a Recipe for its Calculation', *Acta Biotheoretica*, 2002.

39. Smith D.L. et al., 'Ross, Macdonald, and a Theory for the Dynamics and Control of Mosquito-Transmitted Pathogens', *PLOS Pathogens*, 2012.

40. Nájera J.A. et al., 'Some Lessons for the Future from the Global

Malaria Eradication Programme (1955–1969)', *PLOS Medicine*, 2011. A proposal to eradicate smallpox had also been made in 1953, but was met with limited enthusiasm.

41. Background on the reproduction number from: Heesterbeek J.A., 'A Brief History of R0 and a Recipe for its Calculation', *Acta Biotheoretica*, 2002.

42. Abbott S et al. Temporal variation in transmission during the COVID-19 outbreak. CMMID COVID-19 repository. Available from: https://cmmid.github.io/topics/covid19/current-patterns-transmission/global-time-varying-transmission.html.

43. Reproduction number estimates: Fraser C. et al., 'Pandemic potential of a strain of influenza A (H1N1): early findings', *Science*, 2009; WHO Ebola Response Team, 'Ebola Virus Disease in West Africa – The First 9 Months of the Epidemic and Forward Projections', *NEJM*, 2014; Riley S. et al., 'Transmission dynamics of the etiological agent of SARS in Hong Kong', *Science*, 2003; Gani R. and Leach S., 'Transmission potential of smallpox in contemporary populations', *Nature*, 2001; Anderson R.M. and May R.M., *Infectious Diseases of Humans: Dynamics and Control* (Oxford University Press, Oxford, 1992); Guerra F.M. et al., 'The basic reproduction number (R0) of measles: a systematic review', *The Lancet*, 2017.

44. Centers for Disease Control and Prevention, 'Transmission of Measles', 2017. https://www.cdc.gov/measles/transmission/html.

45. Fine P.E.M. and Clarkson J.A., 'Measles in England and Wales – -I: An Analysis of Factors Underlying Seasonal Patterns', *International Journal of Epidemiology*, 1982.

46. 'How Princess Diana changed attitudes to AIDS', BBC News Online, 5 April 2017.

47. May R.M. and Anderson R.M., 'Transmission dynamics of HIV infection', *Nature*, 1987.

48. Eakle R. et al., 'Pre-exposure prophylaxis (PrEP) in an era of stalled HIV prevention: Can it change the game?', *Retrovirology*, 2018.

49. Anderson R.M. and May R.M., *Infectious Diseases of Humans: Dynamics and Control* (Oxford University Press, Oxford, 1992).

50. Fenner F. et al., 'Smallpox and its Eradication', World Health Organization, 1988.

51. Wehrle P.F. et al., 'An Airborne Outbreak of Smallpox in a German Hospital and its Significance with Respect to Other Recent Outbreaks in Europe', *Bulletin of the World Health Organization*, 1970.

52. Woolhouse M.E.J. et al., 'Heterogeneities in the transmission of infectious agents: Implications for the design of control programs', *PNAS*, 1997. The idea built on an earlier observation by nineteenth-century economist Vilfredo Pareto, who'd spotted that 20 per cent of Italians owned 80 per cent of the land.

53. Endo A. et al., 'Estimating the overdispersion in COVID-19 transmission using outbreak sizes outside China', *Wellcome Open Research*, 2020; Adam D. C. et al., 'Clustering and superspreading potential of SARS-CoV-2 infections in Hong Kong', *Nature Med*, 2020.

54. Lloyd-Smith J.O. et al., 'Superspreading and the effect of individual variation on disease emergence', *Nature*, 2005.

55. Worobey M. et al., '1970s and "Patient 0" HIV-1 genomes illuminate early HIV/AIDS history in North America', *Nature*, 2016.

56. Cumming J.G., 'An epidemic resulting from the contamination of ice cream by a typhoid carrier', *Journal of the American Medical Association*, 1917.

57. Bollobas B., 'To Prove and Conjecture: Paul Erdős and His Mathematics', *American Mathematical Monthly*, 1998.

58. Potterat J.J., et al., 'Sexual network structure as an indicator of epidemic phase', *Sexually Transmitted Infections*, 2002.

59. Watts D.J. and Strogatz S.H., 'Collective dynamics of "small-world" networks', *Nature*, 1998.

60. Barabási A.L. and Albert R., 'Emergence of Scaling in Random Networks', *Science*, 1999. A similar idea had emerged in the 1970s, when physicist Derek de Solla Price analysed academic publications. He'd suggested preferential attachment could explain the extreme variation in the number of citations: a paper was more likely to be cited if it was already highly cited. Source: Price D.D.S., 'A General Theory of Bibliometric and Other Cumulative Advantage Processes', *Journal of the American Society for Information Science*, 1976.

61. Liljeros F. et al., 'The web of human sexual contacts', *Nature*, 2001; de Blasio B. et al., 'Preferential attachment in sexual networks', *PNAS*, 2007.

62. Yorke J.A. et al., 'Dynamics and control of the transmission of gonorrhea', *Sexually Transmitted Diseases*, 1978.

63. May R.M. and Anderson R.M., 'The Transmission Dynamics of Human Immunodeficiency Virus (HIV)', *Philosophical Transactions of the Royal Society B*, 1988.

64. Foy B.D. et al., 'Probable Non–Vector-borne Transmission of Zika Virus, Colorado, USA', *Emerging Infectious Diseases*, 2011.

65. Counotte M.J. et al., 'Sexual transmission of Zika virus and other flaviviruses: A living systematic review', *PLOS Medicine*, 2018; Folkers K.M., 'Zika: The Millennials' S.T.D.?', *New York Times*, 20 August 2016.

66. Others independently reached the same conclusion. Sources: Yakob L. et al., 'Low risk of a sexually-transmitted Zika virus outbreak', *The Lancet Infectious Diseases*, 2016; Althaus C.L. and Low N., 'How Relevant Is Sexual Transmission of Zika Virus?' *PLOS Medicine*, 2016.

67. Background on early HIV/AIDS transmission from: Worobey et al. '1970s and "Patient 0" HIV-1 genomes illuminate early HIV/AIDS history in North America', *Nature*, 2016.; McKay R.A., '"Patient Zero": The Absence of a Patient's View of the Early North American AIDS Epidemic', *Bulletin of the History of Medicine*, 2014.

68. This was before the CDC name changed to Centers for Disease Control and Prevention in 1992.

69. McKay R.A. "Patient Zero": The Absence of a Patient's View of the Early North American AIDS Epidemic. Bull Hist Med, 2014.

70. Sapatkin D., 'AIDS: The truth about Patient Zero', *The Philadelphia Inquirer*, 6 May 2013.

71. WHO. Mali case, 'Ebola imported from Guinea: Ebola situation assessment', 10 November 2014.

72. Robert A. et al., 'Determinants of transmission risk during the late stage of the West African Ebola epidemic', *American Journal of Epidemiology*, 2019.

73. Nagel T., 'Moral Luck', 1979.

74. Potterat J.J. et al., 'Gonorrhoea as a Social Disease', *Sexually Transmitted Diseases*, 1985.

75. Potterat J.J., *Seeking The Positives: A Life Spent on the Cutting Edge of Public Health* (Createspace, 2015).

76. Kilikpo Jarwolo J.L., 'The Hurt – and Danger – of Ebola Stigma', ActionAid, 2015.

77. Gregory A. et al., 'Coronavirus: hunt for Patient Zero, Britain's virus spreader', *Sunday Times*, 1 March 2020.

78. Meyer R., Madrigal A., 'Exclusive: The Strongest Evidence Yet That America Is Botching Coronavirus Testing', *The Atlantic*, 6 March 2020; UK testing data available from: https://en.wikipedia.org/wiki/Template:COVID-19_pandemic_data/United_Kingdom_medical_cases

79. Frith J., 'Syphilis – Its Early History and Treatment until Penicillin and the Debate on its Origins', *Journal of Military and Veterans' Health*, 2012.

80. Badcock J., 'Pepe's story: How I survived Spanish flu', BBC News Online, 21 May 2018.

81. Enserink M., 'War Stories', *Science*, 15 March 2013.

82. Lee J-W. and McKibbin W.J., 'Estimating the global economic costs of SARS', from *Learning from SARS: Preparing for the Next Disease Outbreak: Workshop Summary* (National Academies Press, 2004).

83. Haldane A., 'Rethinking the Financial Network', Bank of England, 28 April 2009.

84. Crampton T., 'Battling the spread of SARS, Asian nations escalate travel restrictions', *New York Times*, 12 April 2003. Although travel restrictions were imposed during the outbreak, such restrictions are likely to have had less effect on containment than measures such as case identification and contact tracing. Indeed, WHO did not recommend restrictions during this period: 'World Health Organization. Summary of WHO measures related to international travel', WHO, 24 June 2003.

85. Owens R.E. and Schreft S.L., 'Identifying Credit Crunches', *Contemporary Economic Policy*, 1995.

86. Background and quotes from author interview with Andy Haldane, July 2018.

87. Soramäki K. et al., 'The topology of interbank payment flows', *Federal Reserve Bank of New York Staff Report*, 2006.

88. Gupta S. et al., 'Networks of sexual contacts: implications for the pattern of spread of HIV', *AIDS*, 1989.

89. Haldane A. and May R.M., 'The birds and the bees, and the big banks', *Financial Times*, 20 February 2011.

90. Haldane A., 'Rethinking the Financial Network', Bank of England, 28 April 2009.

91. Buffett W., Letter to the Shareholders of Berkshire Hathaway Inc., 27 February 2009.

92. Keynes J.M., 'The Consequences to the Banks of the Collapse of Money Values', 1931 (from *Essays in Persuasion*).

93. Tavakoli J., Comments on SEC Proposed Rules and Oversight of NRSROs. Letter to Securities and Exchange Commission, 13 February 2007.

94. Arinaminpathy N. et al., 'Size and complexity in model financial systems', *PNAS*, 2012; Caccioli F. et al., 'Stability analysis of financial contagion due to overlapping portfolios', *Journal of Banking & Finance*, 2014; Bardoscia M. et al., 'Pathways towards instability in financial networks', *Nature Communications*, 2017.

95. Haldane A. and May R.M., 'The birds and the bees, and the big banks', *Financial Times*, 20 February 2011.

96. Authers J., 'In a crisis, sometimes you don't tell the whole story', *Financial Times*, 8 September 2018.

97. Arinaminpathy N. et al., 'Size and complexity in model financial systems', *PNAS*, 2012.

98. Independent Commission on Banking. Final Report Recommendations, September 2011.

99. Withers I., 'EU banks spared ringfencing rules imposed on British lenders', *The Telegraph*, 24 October 2017.

100. Bank for International Settlements. Statistical release: 'OTC derivatives statistics at end-June 2018', 31 October 2018.

101. Author interview with Barbara Casu, September 2018.

102. Jenkins P., 'How much of a systemic risk is clearing?' *Financial Times*, 8 January 2018.

103. Battiston S. et al., 'The price of complexity in financial networks', *PNAS*, 2016.

3. The measure of friendship

   1. Background from: Shifman M., *ITEP Lectures in Particle Physics*, arXiv, 1995.
   2. Pais A. J., *Robert Oppenheimer: A Life* (Oxford University Press, 2007).
   3. Goffman W. and Newill V.A., 'Generalization of epidemic theory: An application to the transmission of ideas', *Nature*, 1964. There are some limits to Goffman's analogy, however. In particular, he claimed that the SIR model would be appropriate for the spread of rumours, but others have argued that simple tweaks to the model can produce very different results. For example, in a simple epidemic model, we usually assume people stop being infectious after a period of time, which is reasonable for many diseases. Daryl Daley and David Kendall, two Cambridge mathematicians, have proposed that in a rumour model, spreaders won't necessarily recover naturally; they may only stop spreading the rumour when they meet someone else who's heard the rumour. Source: Daley D.J. and Kendall D.G., 'Epidemics and rumours', *Nature*, 1964.
   4. Landau genius scale. http://www.eoht.info/page/ Landau+genius+scale.
   5. Khalatnikov I.M and Sykes J.B. (eds.), *Landau: The Physicist and the Man: Recollections of L.D. Landau* (Pergamon, 2013).
   6. Bettencourt L.M.A. et al., 'The power of a good idea: Quantitative modeling of the spread of ideas from epidemiological models', *Physica A*, 2006.
   7. Azouly P. et al., 'Does Science Advance One Funeral at a Time?', *National Bureau of Economic Research working paper*, 2015.
   8. Catmull E., 'How Pixar Fosters Collective Creativity', *Harvard Business Review*, September 2008.
   9. Grove J., 'Francis Crick Institute: "gentle anarchy" will fire research', *THE*, 2 September 2016.
   10. Bernstein E.S. and Turban S., 'The impact of the "open"workspace on human collaboration.' *Philosophical Transactions of the Royal Society B*, 2018.
   11. Background and quotes from: 'History of the National Survey

of Sexual Attitudes and Lifestyles'. Witness Seminar held by the
Wellcome Trust Centre for the History of Medicine at UCL,
London, on 14 December 2009.

12. Mercer C.H. et al., 'Changes in sexual attitudes and lifestyles in
Britain through the life course and over time: findings from the
National Surveys of Sexual Attitudes and Lifestyles (Natsal)', *The
Lancet*, 2013.

13. http://www.bbc.co.uk/pandemic.

14. Van Hoang T. et al., 'A systematic review of social contact surveys
to inform transmission models of close contact infections', *BioRxiv*,
2018.

15. Mossong J. et al., 'Social Contacts and Mixing Patterns Relevant
to the Spread of Infectious Diseases', *PLOS Medicine*, 2008;
Kucharski A.J. et al., 'The Contribution of Social Behaviour to
the Transmission of Influenza A in a Human Population', *PLOS
Pathogens*, 2014.

16. Eames K.T.D. et al., 'Measured Dynamic Social Contact Patterns
Explain the Spread of H1N1v Influenza', *PLOS Computational
Biology*, 2012; Eames K.T.D., 'The influence of school holiday
timing on epidemic impact', *Epidemiology and Infection*, 2013;
Baguelin M. et al., 'Vaccination against pandemic influenza A/
H1N1v in England: a real-time economic evaluation', *Vaccine*, 2010.

17. Eames K.T.D., 'The influence of school holiday timing on
epidemic impact', *Epidemiology and Infection*, 2013.

18. Eggo R.M. et al., 'Respiratory virus transmission dynamics
determine timing of asthma exacerbation peaks: Evidence from a
population-level model', *PNAS*, 2016.

19. Kucharski A.J. et al., 'The Contribution of Social Behaviour to
the Transmission of Influenza A in a Human Population', *PLOS
Pathogens*, 2014.

20. Byington C.L. et al., 'Community Surveillance of Respiratory
Viruses Among Families in the Utah Better Identification of
Germs-Longitudinal Viral Epidemiology (BIG-LoVE) Study',
Clinical Infectious Diseases, 2015.

21. Brockmann D. and Helbing D., 'The Hidden Geometry of
Complex, Network-Driven Contagion Phenomena', *Science*, 2013.

22. Gog J.R. et al., 'Spatial Transmission of 2009 Pandemic Influenza in the US', *PLOS Computational Biology*, 2014.
23. Keeling M.J. et al., 'Individual identity and movement networks for disease metapopulations', *PNAS*, 2010.
24. Odlyzko A., 'The forgotten discovery of gravity models and the inefficiency of early railway networks', 2015.
25. Christakis N.A. and Fowler J.H., 'Social contagion theory: examining dynamic social networks and human behavior', *Statistics in Medicine*, 2012.
26. Cohen-Cole E. and Fletcher J.M., 'Detecting implausible social network effects in acne, height, and headaches: longitudinal analysis', *British Medical Journal*, 2008.
27. Lyons R., 'The Spread of Evidence-Poor Medicine via Flawed Social-Network Analysis', *Statistics, Politics, and Policy*, 2011.
28. Norscia I. and Palagi E., 'Yawn Contagion and Empathy in Homo sapiens', *PLOS ONE*, 2011. Note that although it's fairly easy to set up yawn experiments, there can still be challenges with interpreting the results. See: Kapitány R. and Nielsen M., 'Are Yawns really Contagious? A Critique and Quantification of Yawn Contagion', *Adaptive Human Behavior and Physiology*, 2017.
29. Norscia I. et al., 'She more than he: gender bias supports the empathic nature of yawn contagion in Homo sapiens', *Royal Society Open Science*, 2016.
30. Millen A. and Anderson J.R., 'Neither infants nor toddlers catch yawns from their mothers', *Royal Society Biology Letters*, 2010.
31. Holle H. et al., 'Neural basis of contagious itch and why some people are more prone to it'. *PNAS*, 2012; Sy T. et al., 'The Contagious Leader: Impact of the Leader's Mood on the Mood of Group Members, Group Affective Tone, and Group Processes', *Journal of Applied Psychology*, 2005; Johnson S.K., 'Do you feel what I feel? Mood contagion and leadership outcomes', *The Leadership Quarterly*, 2009; Bono J.E. and Ilies R., 'Charisma, positive emotions and mood contagion', *The Leadership Quarterly*, 2006.
32. Sherry D.F. and Galef B.G., 'Cultural Transmission Without Imitation: Milk Bottle Opening by Birds', *Animal Behaviour*, 1984.
33. Background from: Aplin L.M. et al., 'Experimentally induced

innovations lead to persistent culture via conformity in wild birds', *Nature*, 2015. Quotes from author interview with Lucy Aplin, August 2017.

34. Weber M., *Economy and Society* (Bedminster Press Incorporated, New York, 1968).

35. Manski C., 'Identification of Endogenous Social Effects: The Reflection Problem', *Review of Economic Studies*, 1993.

36. Datar A. and Nicosia N., 'Association of Exposure to Communities With Higher Ratios of Obesity With Increased Body Mass Index and Risk of Overweight and Obesity Among Parents and Children' *JAMA Pediatrics*, 2018.

37. Quotes from author interview with Dean Eckles, August 2017.

38. Editorial, 'Epidemiology is a science of high importance', *Nature Communications*, 2018.

39. Background on smoking and cancer from: Howick J. et al., 'The evolution of evidence hierarchies: what can Bradford Hill's "guidelines for causation" contribute?', *Journal of the Royal Society of Medicine*, 2009; Mourant A., 'Why Arthur Mourant Decided To Say "No" To Ronald Fisher', *The Scientist*, 12 December 1988.

40. Background from: Ross R., Memoirs, *With a Full Account of the Great Malaria Problem and its Solution* (London, 1923).

41. Racaniello V., 'Koch's postulates in the 21st century', *Virology Blog*, 22 January 2010.

42. Alice Stewart's obituary, *The Telegraph*, 16 August 2002.

43. Rasmussen S.A. et al., 'Zika Virus and Birth Defects – Reviewing the Evidence for Causality', *NEJM*, 2016.

44. Greene G., *The Woman Who Knew Too Much: Alice Stewart and the Secrets of Radiation* (University of Michigan Press, 2001).

45. Background and quotes from author interview with Nicholas Christakis, June 2018.

46. Snijders T.A.B., 'The Spread of Evidence-Poor Medicine via Flawed Social-Network Analysis', *SOCNET Archives*, 17 June 2011.

47. Granovetter M.S., 'The Strength of Weak Ties', *American Journal of Sociology*, 1973.

48. Dhand A., 'Social networks and risk of delayed hospital arrival after acute stroke', *Nature Communications*, 2019.

49. Background from: Centola D. and Macy M., 'Complex Contagions and the Weakness of Long Ties', *American Journal of Sociology*, 2007; Centola D., *How Behavior Spreads: The Science of Complex Contagions* (Princeton University Press, 2018).

50. Darley J.M. and Latane B., 'Bystander intervention in emergencies: Diffusion of responsibility', *Journal of Personality and Social Psychology*, 1968.

51. Centola D., *How Behavior Spreads: The Science of Complex Contagions* (Princeton University Press, 2018).

52. Coviello L. et al., 'Detecting Emotional Contagion in Massive Social Networks', *PLOS ONE*, 2014; Aral S. and Nicolaides C., 'Exercise contagion in a global social network', *Nature Communications*, 2017.

53. Fleischer D., Executive Summary. The Prop 8 Report, 2010. http://prop8report.lgbtmentoring.org/read-the-report/executive-summary.

54. Background on deep canvassing from: Issenberg S., 'How Do You Change Someone's Mind About Abortion? Tell Them You Had One', *Bloomberg*, 6 October 2014; Resnick B., 'These scientists can prove it's possible to reduce prejudice', *Vox*, 8 April 2016; Bohannon J., 'For real this time: Talking to people about gay and transgender issues can change their prejudices', *Associated Press*, 7 April 2016.

55. Mandel D.R., 'The psychology of Bayesian reasoning', *Frontiers in Psychology*, 2014.

56. Nyhan B. and Reifler J., 'When Corrections Fail: The persistence of political misperceptions', *Political Behavior*, 2010.

57. Wood T. and Porter E., 'The elusive backfire effect: mass attitudes' steadfast factual adherence', *Political Behavior*, 2018.

58. LaCour M.H. and Green D.P., 'When contact changes minds: An experiment on transmission of support for gay equality', *Science*, 2014.

59. Broockman D. and Kalla J., 'Irregularities in LaCour (2014)', Working paper, May 2015.

60. Duran L., 'How to change views on trans people? Just get personal', Take Two®, 7 April 2016.

61. Comment from: Gelman A., 'LaCour and Green 1, This American

Life 0', 16 December 2015. https://statmodeling.stat.columbia.
edu/2015/12/16/lacour-and-green-1-this-american-life-0/

62. Wood T. and Porter E., 'The elusive backfire effect: mass attitudes' steadfast factual adherence', *Political Behavior*, 2018.

63. Weiss R. and Fitzgerald M., 'Edwards, First Lady at Odds on Stem Cells', *Washington Post*, 10 August 2004.

64. Quotes from author interview with Brendan Nyhan, November 2018.

65. Nyhan B. et al., 'Taking Fact-checks Literally But Not Seriously? The Effects of Journalistic Fact-checking on Factual Beliefs and Candidate Favorability', *Political Behavior*, 2019.

66. Example: https://twitter.com/brendannyhan/status/ 859573499333136384.

67. Strudwick P.A., 'Former MP Has Made A Heartfelt Apology For Voting Against Same-Sex Marriage', *BuzzFeed*, 28 March 2017.

68. There's also evidence that people who have changed their mind about a topic, and explain why they've changed their mind, can be more persuasive than a simple one-sided message. Source: Lyons B.A. et al., 'Conversion messages and attitude change: Strong arguments, not costly signals', *Public Understanding of Science*, 2019.

69. Feinberg M. and Willer R., 'From Gulf to Bridge: When Do Moral Arguments Facilitate Political Influence?', *Personality and Social Psychology Bulletin*, 2015.

70. Roghanizad M.M. and Bohns V.K., 'Ask in person: You're less persuasive than you think over email', *Journal of Experimental Social Psychology*, 2016.

71. How J.J. and De Leeuw E.D., 'A comparison of nonresponse in mail, telephone, and face-to-face surveys', *Quality and Quantity*, 1994; Gerber A.S. and Green D.P., 'The Effects of Canvassing, Telephone Calls, and Direct Mail on Voter Turnout: A Field Experiment', *American Political Science Review*, 2000; Okdie B.M. et al., 'Getting to know you: Face-to-face versus online interactions', *Computers in Human Behavior*, 2011.

72. Swire B. et al., 'The role of familiarity in correcting inaccurate information', *Journal of Experimental Psychology Learning Memory and Cognition*, 2017.

73. Quotes from author interview with Briony Swire-Thompson, July 2018.

74. Broockman D. and Kalla J., 'Durably reducing transphobia: A field experiment on door-to-door canvassing', *Science*, 2016.

4. Something in the air

1. Background and quotes from author interview with Gary Slutkin, April 2018.

2. Statistics from: Bentle K. et al., '39,000 homicides: Retracing 60 years of murder in Chicago', *Chicago Tribune*, 9 January 2018; Illinois State Fact Sheet. National Injury and Violence Prevention Resource Center, 2015.

3. Slutkin G., 'Treatment of violence as an epidemic disease', In: Fine P. et al. John Snow's legacy: epidemiology without borders. *The Lancet*, 2013.

4. Background on John Snow's work on cholera from: Snow J., *On the mode of communication of cholera*. (London, 1855); Tulodziecki D., 'A case study in explanatory power: John Snow's conclusions about the pathology and transmission of cholera', *Studies in History and Philosophy of Biological and Biomedical Sciences*, 2011; Hempel S., 'John Snow', *The Lancet*, 2013; Brody H. et al., 'Map-making and myth-making in Broad Street: the London cholera epidemic, 1854', *The Lancet*, 2000.

5. Reason for abstraction: Seuphor M., *Piet Mondrian: Life and Work* (Abrams, New York, 1956); Tate Modern, 'Five ways to look at Malevich's Black Square', https://www.tate.org.uk/art/artists/kazimir-malevich-1561/five-ways-look-malevichs-black-square.

6. Background on cholera: Locher W.G., 'Max von Pettenkofer (1818–1901) as a Pioneer of Modern Hygiene and Preventive Medicine', *Environmental Health and Preventive Medicine*, 2007; Morabia A., 'Epidemiologic Interactions, Complexity, and the Lonesome Death of Max von Pettenkofer,' *American Journal of Epidemiology*, 2007.

7. García-Moreno C. et al., 'WHO Multi-country Study on Women's Health and Domestic Violence against Women', *World Health Organization*, 2005.

8. Quotes from author interview with Charlotte Watts, May 2018.

9. Background on factors influencing contagion of violence: Patel

D.M. et al., *Contagion of Violence: Workshop Summary* (National Academies Press, 2012).

10. Gould M.S. et al., 'Suicide Clusters: A Critical Review', *Suicide and Life-Threatening Behavior*, 1989.

11. Cheng Q. et al., 'Suicide Contagion: A Systematic Review of Definitions and Research Utility', *PLOS ONE*, 2014.

12. Phillips D.P., 'The Influence of Suggestion on Suicide: Substantive and Theoretical Implications of the Werther Effect', *American Sociological Review*, 1974.

13. WHO. 'Is responsible and deglamourized media reporting effective in reducing deaths from suicide, suicide attempts and acts of self-harm?', 2015. https://www.who.int.

14. Fink D.S. et al., 'Increase in suicides the months after the death of Robin Williams in the US', *PLOS ONE*, 2018.

15. Towers S. et al., 'Contagion in Mass Killings and School Shootings', *PLOS ONE*, 2015.

16. Brent D.A. et al., 'An Outbreak of Suicide and Suicidal Behavior in a High School', *Journal of the American Academy of Child and Adolescent Psychiatry*, 1989.

17. Aufrichtig A. et al., 'Want to fix gun violence in America? Go local', *The Guardian*, 9 January 2017.

18. Quotes from author interview with Charlie Ransford, April 2018.

19. Confino J., 'Guardian-supported Malawi sex workers' project secures funding from Comic Relief', *The Guardian*, 9 June 2010.

20. Bremer S., '10 Shot, 2 Fatally, at Vigil on Chicago's Southwest Side', *NBC Chicago*, 7 May 2017.

21. Tracy M. et al., 'The Transmission of Gun and Other Weapon-Involved Violence Within Social Networks', *Epidemiologic Reviews*, 2016.

22. Green B. et al., 'Modeling Contagion Through Social Networks to Explain and Predict Gunshot Violence in Chicago, 2006 to 2014', *JAMA Internal Medicine*, 2017.

23. Fitting a negative binomial offspring distribution to the cluster size distribution from Green et al., I obtained a maximum likelihood estimate for the dispersion parameter $k=0.096$. (Method from: Blumberg S. and Lloyd-Smith J.O., *PLOS Computational Biology*,

2013.) For context, MERS-CoV had R=0.63 and k=0.25 (from: Kucharski A.J. and Althaus C.L., 'The role of superspreading in Middle East respiratory syndrome coronavirus (MERS-CoV) transmission', *Eurosurveillance*, 2015).

24. Fenner F. et al., *Smallpox and its Eradication* (World Health Organization, Geneva, 1988).

25. Ganyani T. et al., 'Estimating the generation interval for coronavirus disease (COVID-19) based on symptom onset data, March 2020.' *Eurosurveillance*, 2020.

26. Evaluations of violence interruption methods: Skogan W.G. et al., 'Evaluation of CeaseFire-Chicago', U.S. Department of Justice report, March 2009; Webster D.W. et al., 'Evaluation of Baltimore's Safe Streets Program', Johns Hopkins report, January 2012; Thomas R. et al., 'Investing in Intervention: The Critical Role of State-Level Support in Breaking the Cycle of Urban Gun Violence', Giffords Law Center report, 2017.

27. Examples of criticism of Cure Violence: Page C., 'The doctor who predicted Chicago's homicide epidemic', *Chicago Tribune*, 30 December 2016; 'We need answers on anti-violence program', *Chicago Sun Times*, 1 July 2014.

28. Patel D.M. et al., *Contagion of Violence: Workshop Summary* (National Academies Press, 2012).

29. Background from: Seenan G., 'Scotland has second highest murder rate in Europe', *The Guardian*, 26 September 2005; Henley J., 'Karyn McCluskey: the woman who took on Glasgow's gangs', *The Guardian*, 19 December 2011; Ross P., 'No mean citizens: The success behind Glasgow's VRU', *The Scotsman*, 24 November 2014; Geoghegan P., 'Glasgow smiles: how the city halved its murders by "caring people into change"', *The Guardian*, 6 April 2015; '10 Year Strategic Plan', Scottish Violence Reduction Unit, 2017.

30. Adam K., 'Glasgow was once the "murder capital of Europe". Now it's a model for cutting crime', *Washington Post*, 27 October 2018.

31. Formal evaluations are not available for all aspects of the VRU programme, but some parts have been evaluated: Williams D.J. et al., 'Addressing gang-related violence in Glasgow: A preliminary

pragmatic quasi-experimental evaluation of the Community
Initiative to Reduce Violence (CIRV)', *Aggression and Violent
Behavior,* 2014; Goodall C. et al., 'Navigator: A Tale of Two Cities',
12 Month Report, 2017.

32. 'Mayor launches new public health approach to tackling serious
violence', London City Hall press release, 19 September 2018;
Bulman M., 'Woman who helped dramatically reduce youth
murders in Scotland urges London to treat violence as a "disease"',
The Independent, 5 April 2018.

33. Background on Nightingale's Crimea work from: Gill C.J. and Gill
G.C., 'Nightingale in Scutari: Her Legacy Reexamined', *Clinical
Infectious Diseases,* 2005; Nightingale F., *Notes on Matters Affecting
the Health, Efficiency, and Hospital Administration of the British Army:
Founded Chiefly on the Experience of the Late War* (London, 1858);
Magnello M.E., 'Victorian statistical graphics and the iconography
of Florence Nightingale's polar area graph', *Journal of the British
Society for the History of Mathematics Bulletin,* 2012.

34. Nelson S. and Rafferty A.M., *Notes on Nightingale: The Influence and
Legacy of a Nursing Icon* (Cornell University Press, 2012).

35. Background on Farr from: Lilienfeld D.E., 'Celebration: William
Farr (1807–1883) – an appreciation on the 200th anniversary of his
birth', *International Journal of Epidemiology,* 2007; Humphreys N.A.,
'Vital statistics: a memorial volume of selections from the reports
and writings of William Farr', *The Sanitary Institute of Great Britain,*
1885.

36. Nightingale F., *A Contribution to the Sanitary History of the British
Army During the Late War with Russia* (London, 1859).

37. Quoted in: Diamond M. and Stone M., 'Nightingale on Quetelet',
Journal of the Royal Statistical Society A, 1981.

38. Cook E., *The Life of Florence Nightingale* (London, 1913).

39. Quoted in: MacDonald L., *Florence Nightingale on Society and
Politics, Philosophy, Science, Education and Literature* (Wilfrid Laurier
University Press, 2003).

40. Pearson K., *The Life, Letters and Labours of Francis Galton*
(Cambridge University Press, London, 1914).

41. Patel D.M. et al., *Contagion of Violence: Workshop Summary* (National Academies Press, 2012).

42. Statistics from: Grinshteyn E. and Hemenway D., 'Violent Death Rates: The US Compared with Other High-income OECD Countries, 2010', *The American Journal of Medicine*, 2016; Koerth-Baker M., 'Mass Shootings Are A Bad Way To Understand Gun Violence', *Five Thirty Eight*, 3 October 2017.

43. Background from: Thompson B., 'The Science of Violence', *Washington Post*, 29 March 1998; Wilkinson F., 'Gunning for Guns', *Rolling Stone*, 9 December 1993.

44. Cillizza C., 'President Obama's amazingly emotional speech on gun control', *Washington Post*, 5 January 2016.

45. Borger J., 'The Guardian profile: Ralph Nader', *The Guardian*, 22 October 2004.

46. Background from: Jensen C., '50 Years Ago, "Unsafe at Any Speed" Shook the Auto World', *New York Times*, 26 November 2015.

47. Kelly K., 'Car Safety Initially Considered "Undesirable" by Manufacturers, the Government and Consumers', *Huffington Post*, 4 December 2012.

48. Frankel T.C., 'Their 1996 clash shaped the gun debate for years. Now they want to reshape it', *Washington Post*, 30 December 2015.

49. Kates D.B. et al., 'Public Health Pot Shots', *Reason*, April 1997.

50. Turvill J.L. et al., 'Change in occurrence of paracetamol overdose in UK after introduction of blister packs', *The Lancet*, 2000; Hawton K. et al., 'Long term effect of reduced pack sizes of paracetamol on poisoning deaths and liver transplant activity in England and Wales: interrupted time series analyses', *British Medical Journal*, 2013.

51. Dickey J. and Rosenberg M., 'We won't know the cause of gun violence until we look for it', *Washington Post*, 27 July 2012.

52. Background and quotes from author interview with Toby Davies, August 2017.

53. Davies T.P. et al., 'A mathematical model of the London riots and their policing', *Scientific Reports*, 2013.

54. Example: Myers P., 'Staying streetwise', *Reuters*, 8 September 2011.

55. Quoted in: De Castella T. and McClatchey C., 'UK riots: What turns people into looters?', BBC News Online. 9 August 2011.

56. Granovetter M., 'Threshold Models of Collective Behavior', *American Journal of Sociology*, 1978.

57. Background from: Johnson N.F. et al., 'New online ecology of adversarial aggregates: ISIS and beyond', *Science*, 2016; Wolchover N., 'A Physicist Who Models ISIS and the Alt-Right', *Quanta Magazine*, 23 August 2017.

58. Bohorquez J.C. et al., 'Common ecology quantifies human insurgency', *Nature*, 2009.

59. Belluck P., 'Fighting ISIS With an Algorithm, Physicists Try to Predict Attacks', *New York Times*, 16 June 2016.

60. Timeline: 'How The Anthrax Terror Unfolded', National Public Radio (NPR), 15 February 2011.

61. Cooper B., 'Poxy models and rash decisions', *PNAS*, 2006; Meltzer M.I. et al., 'Modeling Potential Responses to Smallpox as a Bioterrorist Weapon', *Emerging Infectious Diseases*, 2001.

62. I've seen the toy train example used in a few fields (e.g. by Emanuel Derman in finance), but particular credit to my old colleague Ken Eames here, who used it very effectively in disease modelling lectures.

63. Meltzer M.I. et al., 'Estimating the Future Number of Cases in the Ebola Epidemic – Liberia and Sierra Leone, 2014–2015', *Morbidity and Mortality Weekly Report*, 2014.

64. The CDC exponential model estimated around a three-fold increase per month. Therefore a prediction three additional months ahead would have estimated 27-fold more cases than the January value. (The combined population of SL, Liberia and Guinea was around 24 million.)

65. 'Expert reaction to CDC estimates of numbers of future Ebola cases', *Science Media Centre*, 24 September 2014.

66. Background from: Hughes M., 'Developers wish people would remember what a big deal Y2K bug was', *The Next Web*, 26 October 2017; Schofield J., 'Money we spent', *The Guardian*, 5 January 2000.

67. https://twitter.com/JoanneLiu_MSF/status/952834207667097600.

68. In the CDC analysis, cases were scaled up by a factor of 2.5 to

account for under-reporting. If we apply the same scaling to the reported cases, this suggests there were around 75,000 infections in reality, a difference of 1.33 million from the CDC prediction. The suggestion that the CDC model with interventions could explain outbreak comes from: Frieden T.R. and Damon I.K., 'Ebola in West Africa – CDC's Role in Epidemic Detection, Control, and Prevention', *Emerging Infectious Diseases*, 2015.

69. Onishi N., 'Empty Ebola Clinics in Liberia Are Seen as Misstep in U.S. Relief Effort', *New York Times*, 2015.

70. Kucharski A.J. et al., 'Measuring the impact of Ebola control measures in Sierra Leone', *PNAS*, 2015.

71. Camacho A. et al., 'Potential for large outbreaks of Ebola virus disease', *Epidemics*, 2014.

72. Heymann D.L., 'Ebola: transforming fear into appropriate action', *The Lancet*, 2017.

73. Widely attributed, but no clear primary source.

74. By early December, the average reporting delay was 2–3 days. Source: Finger F. et al., 'Real-time analysis of the diphtheria outbreak in forcibly displaced Myanmar nationals in Bangladesh', *BMC Medicine*, 2019.

75. Statistics from: Katz J. and Sanger-Katz M., '"The Numbers Are So Staggering." Overdose Deaths Set a Record Last Year', *New York Times*, 29 November 2018; Ahmad F.B. et al., 'Provisional drug overdose death counts', National Center for Health Statistics, 2018; Felter C., 'The U.S. Opioid Epidemic', Council on Foreign Relations, 26 December 2017; 'Opioid painkillers "must carry prominent warnings"'. BBC News Online, 28 April 2019.

76. Goodnough A., Katz J. and Sanger-Katz M., 'Drug Overdose Deaths Drop in U.S. for First Time Since 1990', *New York Times*, 17 July 2019.

77. Background and quotes about opioid crisis analysis from author interview with Rosalie Liccardo Pacula, May 2018. Additional details from: Pacula R.L., Testimony presented before the House Appropriations Committee, Subcommittee on Labor, Health and Human Services, Education, and Related Agencies on April 5, 2017.

78. Exponential increase in death rate from 11 per 100,000 in 1979 to 137

per 100,000 in 2015, implying doubling time $= 36/\log_2(137/11) = 10$ years.

79. Jalal H., 'Changing dynamics of the drug overdose epidemic in the United States from 1979 through 2016', *Science*, 2018.

80. Mars S.G. '"Every 'never' I ever said came true": transitions from opioid pills to heroin injecting', *International Journal of Drug Policy*, 2014.

81. TCR Staff, 'America "Can't Arrest Its Way Out of the Opioid Epidemic"', *The Crime Report*, 16 February 2018.

82. Lum K. and Isaac W., 'To predict and serve?' *Significance*, 7 October 2016.

83. Quotes from author interview with Kristian Lum, January 2018.

84. Perry W.L. et al., 'Predictive Policing', RAND Corporation Report, 2013.

85. Whitty C.J.M., 'What makes an academic paper useful for health policy?', *BMC Medicine*, 2015.

86. Dumke M. and Main F., 'A look inside the watch list Chicago police fought to keep secret', *Associated Press*, 18 June 2017.

87. Background on SSL algorithm: Posadas B., 'How strategic is Chicago's "Strategic Subjects List"? Upturn investigates', *Medium*, 22 June 2017; Asher J. and Arthur R., 'Inside the Algorithm That Tries to Predict Gun Violence in Chicago', *New York Times*, 13 June 2017; Kunichoff Y. and Sier P., 'The Contradictions of Chicago Police's Secretive List', *Chicago Magazine*, 21 August 2017.

88. According to Posadas (*Medium*, 2017), proportion high risk: $287,404/398,684 = 0.72$. 88,592 of these (31 per cent) have never been arrested or a victim of crime.

89. Hemenway D., *While We Were Sleeping: Success Stories in Injury and Violence Prevention*, (University of California Press, 2009).

90. Background on broken windows approach: Kelling G.L. and Wilson J.Q., 'Broken Windows', *The Atlantic*, March 1982; Harcourt B.E. and Ludwig J., 'Broken Windows: New Evidence from New York City and a Five-City Social Experiment', *University of Chicago Law Review*, 2005.

91. Childress S., 'The Problem with "Broken Windows" Policing', Public Broadcasting Service, 28 June 2016.

92. Keizer K. et al., 'The Spreading of Disorder', *Science*, 2008.
93. Keizer K. et al., 'The Importance of Demonstratively Restoring Order', *PLOS ONE*, 2013.
94. Tcherni-Buzzeo M., 'The "Great American Crime Decline": Possible explanations', In Krohn M.D. et al., *Handbook on Crime and Deviance*, 2nd edition, (Springer, New York 2019).
95. Alternative hypotheses for decline, and accompanying criticism: Levitt S.D., 'Understanding Why Crime Fell in the 1990s: Four Factors that Explain the Decline and Six that Do Not', *Journal of Economic Perspectives*, 2004; Nevin R., 'How Lead Exposure Relates to Temporal Changes in IQ, Violent Crime, and Unwed Pregnancy', *Environmental Research Section A*, 2000; Foote C.L. and Goetz C.F., 'The Impact of Legalized Abortion on Crime: Comment', *Quarterly Journal of Economics*, 2008; Casciani D., 'Did removing lead from petrol spark a decline in crime?', BBC News Online, 21 April 2014.
96. Author interview with Melissa Tracy, August 2018.
97. Lowrey A., 'True Crime Costs', *Slate*, 21 October 2010.

5. Going viral

1. Background on Buzzfeed from: Peretti J., 'My Nike Media Adventure', *The Nation*, 9 April 2001; Email correspondence with customer service representatives at Nike iD. http://www.yorku.ca/dzwick/niked.html Accessed: January 2018; Salmon F., 'BuzzFeed's Jonah Peretti Goes Long', *Fusion*, 11 June 2014; Lagorio-Chafkin C., 'The Humble Origins of Buzzfeed', *Inc.*, 3 March 2014; Rice A., 'Does BuzzFeed Know the Secret?', *New York Magazine*, 7 April 2013.
2. Peretti J., 'My Nike Media Adventure', *The Nation*, 9 April 2001.
3. Background and quotes from author interview with Duncan Watts, February 2018. There is also a more detailed discussion of this research in: Watts D., *Everything is Obvious: Why Common Sense is Nonsense* (Atlantic Books, 2011).
4. Milgram S., 'The small-world problem', *Psychology Today*, 1967.
5. Dodds P.S. et al., 'An Experimental Study of Search in Global Social Networks', *Science*, 2003.
6. Bakshy E. et al., 'Everyone's an Influencer: Quantifying Influence

on Twitter', *Proceedings of the Fourth ACM International Conference on Web Search and Data Mining (WSDM'11)*, 2011.

7. Aral S. and Walker D., 'Identifying Influential and Susceptible Members of Social Networks', *Science*, 2012.

8. Aral S. and Dillon P., 'Social influence maximization under empirical influence models', *Nature Human Behaviour*, 2018.

9. Data from: Ugander J. et al., 'The Anatomy of the Facebook Social Graph', *arXiv*, 2011; Kim D.A. et al., 'Social network targeting to maximise population behaviour change: a cluster randomised controlled trial', *The Lancet*, 2015; Newman M.E., 'Assortative mixing in networks', *Physical Review Letters*, 2002; Apicella C.L. et al., 'Social networks and cooperation in hunter-gatherers', *Nature*, 2012.

10. Conclusion supported by: Aral S. and Dillon P., *Nature Human Behaviour*, 2018; Bakshy E. et al., *WSDM*, 2011; Kim D.A. et al., *The Lancet*, 2015.

11. Buckee C.O.F. et al., 'The effects of host contact network structure on pathogen diversity and strain structure', *PNAS*, 2004; Kucharski A., 'Study epidemiology of fake news', *Nature*, 2016.

12. Bessi A. et al., 'Science vs Conspiracy: Collective Narratives in the Age of Misinformation', *PLOS ONE*, 2015; Garimella K. et al., 'Political Discourse on Social Media: Echo Chambers, Gatekeepers, and the Price of Bipartisanship', *Proceedings of the World Wide Web Conference 2018*, 2018.

13. Background from: Goldacre B., *Bad Science* (Fourth Estate, 2008); The Editors of The Lancet, 'Retraction – Ileal-lymphoid-nodular hyperplasia, non-specific colitis, and pervasive developmental disorder in children', *The Lancet*, 2010.

14. Finnegan G., 'Rise in vaccine hesitancy related to pursuit of purity', *Horizon Magazine*, 26 April 2018; Larson H.J., 'Maternal immunization: The new "normal" (or it should be)', *Vaccine*, 2015; Larson H.J. et al., 'Tracking the global spread of vaccine sentiments: The global response to Japan's suspension of its HPV vaccine recommendation', *Human Vaccines & Immunotherapeutics*, 2014.

15. Background on variation from: 'Variation – an overview', *ScienceDirect Topics*, 2018.

16. Voltaire., 'Letter XI' from *Letters on the English*. (1734).

17. Background on Bernoulli's work from: Dietz K. and Heesterbeek J.A.P., 'Daniel Bernoulli's epidemiological model revisited', *Mathematical Biosciences*, 2002; Colombo C. and Diamanti M., 'The smallpox vaccine: the dispute between Bernoulli and d'Alembert and the calculus of probabilities', *Lettera Matematica International*, 2015.

18. There is a large literature on MMR and measles vaccine safety and efficacy, e.g. Smeeth L. et al., 'MMR vaccination and pervasive developmental disorders: a case-control study', *The Lancet*, 2004; A. Hviid, J.V. Hansen, M. Frisch, et al., 'Measles, Mumps, Rubella Vaccination and Autism: A Nationwide Cohort Study', *Annals of Internal Medicine*, 2019; LeBaron C.W. et al., 'Persistence of Measles Antibodies After 2 Doses of Measles Vaccine in a Postelimination Environment', *JAMA Pediatrics*, 2007.

19. Wellcome Global Monitor 2018, 19 June 2019.

20. Finnegan G., 'Rise in vaccine hesitancy related to pursuit of purity', *Horizon Magazine*, 26 April 2018.

21. Funk S. et al., 'Combining serological and contact data to derive target immunity levels for achieving and maintaining measles elimination', *BioRxiv*, 2019.

22. 'Measles: Europe sees record number of cases and 37 deaths so far this year', *British Medical Journal*, 2018.

23. Bakshy E. et al., 'Exposure to ideologically diverse news and opinion on Facebook', *Science*, 2015; Tufekci Z., 'How Facebook's Algorithm Suppresses Content Diversity (Modestly) and How the Newsfeed Rules Your Clicks', *Medium*, 7 May 2015.

24. Flaxman S. et al., 'Filter bubbles, echo chambers and online news consumption', *Public Opinion Quarterly*, 2016.

25. Bail C.A. et al., 'Exposure to opposing views on social media can increase political polarization', *PNAS*, 2018.

26. Duggan M. and Smith A., 'The Political Environment on Social Media', Pew Research Center, 2016.

27. boyd dm., 'Taken Out of Context: American Teen Sociality in

Networked Publics', University of California, Berkeley PhD
Dissertation, 2008.

28. Early example: 'Dead pet UL?' Posted on alt.folklore.urban, 10 July
1992.

29. Letter to Étienne Noël Damilaville, 16 May 1767.

30. Suler J., 'The Online Disinhibition Effect', *Cyberpsychology and
Behavior*, 2004.

31. Cheng J. et al., 'Antisocial Behavior in Online Discussion
Communities', *Association for the Advancment of Artificial Intelligence*,
2015; Cheng J. et al., 'Anyone Can Become a Troll: Causes of
Trolling Behavior in Online Discussions', Computer-Supported
Cooperative Work, 2017.

32. Background on Facebook study from: Kramer A.D.I. et al.,
'Experimental evidence of massive-scale emotional contagion
through social networks', *PNAS*, 2014; D'Onfro J., 'Facebook
Researcher Responds To Backlash Against "Creepy" Mood
Manipulation Study', *Insider*, 29 June 2014.

33. Griffin A., 'Facebook manipulated users' moods in secret
experiment', *The Independent*, 29 June 2014; Arthur C., 'Facebook
emotion study breached ethical guidelines, researchers say', *The
Guardian*, 30 June 2014.

34. Examples: Raine R. et al., 'A national cluster-randomised
controlled trial to examine the effect of enhanced reminders on
the socioeconomic gradient in uptake in bowel cancer screening',
British Journal of Cancer, 2016; Kitchener H.C. et al., 'A cluster
randomised trial of strategies to increase cervical screening uptake
at first invitation (STRATEGIC)', *Health Technology Assessment*,
2016. It's worth noting that despite their widespread use, the
concept of randomised experiments (often called 'A/B tests')
seems to make many people uncomfortable – even if the individual
options are innocuous and the study is ethically designed. One
2019 study found that 'people frequently rate A/B tests designed
to establish the comparative effectiveness of two policies or
treatments as inappropriate even when universally implementing
either A or B, untested, is seen as appropriate'. Source: Meyer M.N.

et al., 'Objecting to experiments that compare two unobjectionable policies or treatments', *PNAS*, 2019.

35. Berger J. and Milkman K.L., 'What Makes online Content Viral?', *Journal of Marketing Research*, 2011.

36. Heath C. et al., 'Emotional selection in memes: the case of urban legends', *Journal of Personality and Social Psychology*, 2001.

37. Tufekci Z., 'YouTube, the Great Radicalizer', *New York Times*, 10 March 2018.

38. Baquero F. et al., 'Ecology and evolution of antibiotic resistance', *Environmental Microbiology Reports*, 2009.

39. Background from: De Domenico M. et al., 'The Anatomy of a Scientific Rumor', *Scientific Reports*, 2013.

40. Goel S. et al., 'The Structural Virality of Online Diffusion', *Management Science*, 2016.

41. Goel S. et al., 'The Structure of Online Diffusion Networks', *EC'12 Proceedings of the 13th ACM Conference on Electronic Commerce*, 2012; Tatar A. et al., 'A survey on predicting the popularity of web content', *Journal of Internet Services and Applications*, 2014.

42. Watts D.J. et al., 'Viral Marketing for the Real World', *Harvard Business Review*, 2007.

43. Method from: Blumberg S. and Lloyd-Smith J.O., *PLOS Computational Biology*, 2013. This calculation works even if there is potential for superspreading events.

44. Chowell G. et al., 'Transmission potential of influenza A/H7N9, February to May 2013, China', *BMC Medicine*, 2013.

45. Watts D.J. et al., 'Viral Marketing for the Real World', *Harvard Business Review*, 2007. Note that technical issues with the e-mail campaign may have artificially reduced the reproduction number for Tide to some extent.

46. Breban R. et al., 'Interhuman transmissibility of Middle East respiratory syndrome coronavirus: estimation of pandemic risk', *The Lancet*, 2013.

47. Geoghegan J.L. et al., 'Virological factors that increase the transmissibility of emerging human viruses', *PNAS*, 2016.

48. García-Sastre A., 'Influenza Virus Receptor Specificity', *American Journal of Pathology*, 2010.

49. Adamic L.A. et al., 'Information Evolution in Social Networks', *Proceedings of the Ninth ACM International Conference on Web Search and Data Mining (WSDM'16)*, 2016.

50. Cheng J. et al., 'Do Diffusion Protocols Govern Cascade Growth?', *AAAI Publications*, 2018.

51. Background on early BuzzFeed transmission: Rice A., 'Does BuzzFeed Know the Secret?', *New York Magazine*, 7 April 2013.

52. Watts D.J. et al., 'Viral Marketing for the Real World', *Harvard Business Review*, 2007. For ease of reading, the shorthand '<' has been replaced by 'less than' in the text.

53. Guardian Datablog, 'Who are the most social publishers on the web?', *The Guardian Online*, 3 October 2013.

54. Salmon F., 'BuzzFeed's Jonah Peretti Goes Long', *Fusion*, 11 June 2014.

55. Martin T. et al., 'Exploring Limits to Prediction in Complex Social Systems', *Proceedings of the 25th International Conference on World Wide Web*, 2016.

56. Shulman B. et al., 'Predictability of Popularity: Gaps between Prediction and Understanding', *International Conference on Web and Social Media*, 2016.

57. Cheng J. et al., 'Can cascades be predicted?', *Proceedings of the 23rd International Conference on World Wide Web*, 2014.

58. Yucesoy B. et al., 'Success in books: a big data approach to bestsellers', *EPJ Data Science*, 2018.

59. McMahon V., '#Neknominate girl's shame: I'm sorry for drinking a goldfish', *Irish Mirror*, 5 February 2014.

60. Many Neknomination videos can be seen on YouTube; Fricker M., 'RSPCA hunt yob who downed NekNomination cocktail containing cider, eggs, battery fluid, urine and THREE goldfish', *Mirror*, 5 February 2014.

61. Example coverage: Fishwick C., 'NekNominate: should Facebook ban the controversial drinking game?', *The Guardian*, 11 February 2014; '"Neknomination": Facebook ignores calls for ban after two deaths', *Evening Standard*, 3 February 2014.

62. More or Less: 'Neknomination Outbreak', BBC World Service Online, 22 February 2014.

63. Kucharski A.J., 'Modelling the transmission dynamics of online social contagion', *arXiv*, 2016.

64. Researchers at the University of Warwick found a similar level of predictability. Based on the dynamics of neknomination, they correctly forecast the four-week duration of the ice bucket challenge shortly after it emerged a few months later. Sprague D.A. and House T., 'Evidence for complex contagion models of social contagion from observational data', *PLOS ONE*, 2017.

65. Cheng J. et al., 'Do Cascades Recur?', *Proceedings of the 25th International Conference on World Wide Web*, 2016.

66. Crane R. and Sornette D., 'Robust dynamic classes revealed by measuring the response function of a social system', *PNAS*, 2008.

67. Tan C. et al., 'Lost in Propagation? Unfolding News Cycles from the Source', *Association for the Advancement of Artificial Intelligence*, 2016; Tatar A. et al., 'A survey on predicting the popularity of web content', *Journal of Internet Services and Applications*, 2014.

68. Vosoughi S. et al., 'The spread of true and false news online', *Science*, 2018.

69. Examples from: Romero D.M., 'Differences in the Mechanics of Information Diffusion Across Topics: Idioms, Political Hashtags, and Complex Contagion on Twitter', *Proceedings of the 20th International Conference on World Wide Web*, 2011; State B. and Adamic L.A., 'The Diffusion of Support in an Online Social Movement: Evidence from the Adoption of Equal-Sign Profile Pictures', *Proceedings of the 18th ACM Conference on Computer Supported Cooperative Work & Social Computing*, 2015; Guilbeault D. et al., 'Complex Contagions: A Decade in Review', in Lehmann S. and Ahn Y. (eds.), *Spreading Dynamics in Social Systems* (Springer Nature, 2018).

70. Weng L. et al., 'Virality Prediction and Community Structure in Social Networks', *Scientific Reports*, 2013.

71. Centola D., *How Behavior Spreads: The Science of Complex Contagions* (Princeton University Press, 2018).

72. Anderson C., 'The End of Theory: The Data Deluge Makes the Scientific Method Obsolete', *Wired*, 23 June 2008.

73. 'Big Data, for better or worse: 90 per cent of world's data generated over last two years', *Science Daily*, 22 May 2013.

74. Widely attributed to Goodhart in this form. Original statement: 'Any observed statistical regularity will tend to collapse once pressure is placed upon it for control purposes'. Goodhart C., 'Problems of Monetary Management: The U.K. Experience', in Courakis, A. S. (ed.), *Inflation, Depression, and Economic Policy in the West* (Springer 1981).

75. Small J.P., *Wax Tablets of the Mind: Cognitive Studies of Memory and Literacy in Classical Antiquity* (Routledge, 1997).

76. Lewis K. et al., 'The Structure of Online Activism', *Sociological Science*, 2014.

77. Gabielkov M. et al., 'Social Clicks: What and Who Gets Read on Twitter?', ACM SIGMETRICS, 2016.

78. Quotes from author interview with Dean Eckles, August 2017.

79. Widely attributed, but no clear primary source.

80. One common example of ad tracking is the Facebook Pixel. Source: 'Conversion Tracking', Facebook for Developers, 2019. https://developers.facebook.com/docs/facebook-pixel

81. Timeline from: Lederer B., '200 Milliseconds: The Life of a Programmatic RTB Ad Impression', Shelly Palmer, 9 June 2014.

82. Nsubuga J., 'Conservative MP Gavin Barwell in "date Arab girls" Twitter gaffe', *Metro*, 18 March 2013.

83. Albright J., 'Who Hacked the Election? Ad Tech did. Through "Fake News," Identify Resolution and Hyper-Personalization', *Medium*, 30 July 2017.

84. Facebook ad revenue per user in the US and Canada was $30 in Q1 2019, which would suggest $120 per annum. If users are worth 60 per cent less without browser data, it implies average data value of (at least) $120 x 0.6 = $72. Estimates from: Facebook Q1 2019 Results, http://investor.fb.com; Johnson G.A. et al., 'Consumer Privacy Choice in Online Advertising: Who Opts Out and at What Cost to Industry?', *Simon Business School Working paper*, 2017; Leswing K., Apple makes billions from Google's dominance in search – and it's a bigger business than iCloud or Apple Music',

Business Insider, 29 September 2018; Bell K., 'iPhone's user base to surpass 1 billion units by 2019', *Cult of Mac*, 8 February 2017.

85. Pandey E. and Parker S., 'Facebook was designed to exploit human "vulnerability"', *Axios*, 9 November 2017.

86. Kafka P., 'Amazon? HBO? Netflix thinks its real competitor is... sleep', *Vox*, 17 April 2017.

87. Background on design from: Harris T., 'How Technology is Hijacking Your Mind – from a Magician and Google Design Ethicist', *Medium*, 18 May 2016.

88. Bajarin B., 'Apple's Penchant for Consumer Security', *Tech.pinions*, 18 April 2016.

89. Pandey E. and Parker S., 'Facebook was designed to exploit human "vulnerability"', *Axios*, 9 November 2017.

90. Although now a central feature of social media, the 'like' button originated in a very different online era. Source: Locke M., 'How Likes Went Bad', *Medium*, 25 April 2018.

91. Lewis P. '"Our minds can be hijacked": the tech insiders who fear a smartphone dystopia', *Guardian*, 6 October 2017.

92. 'Who can see the comments on my Moments posts?', WeChat Help Center, October 2018.

93. Background on censorship from: King G. et al., 'Reverse-engineering censorship in China: Randomized experimentation and participant observation', *Science*, 2014; Tucker J., 'This explains how social media can both weaken – and strengthen – democracy', *Washington Post*, 6 December 2017.

94. Das S. and Kramer A., *Self-Censorship on Facebook*, AAAI, 2013.

95. Davidsen C., 'You Are Not a Target', 7 June 2015. Full video: https://www.youtube.com/watch?v=LGiiQUMaShw&feature=youtu.be

96. Issenberg S., 'How Obama's Team Used Big Data to Rally Voters', *MIT Technology Review*, 19 December 2012.

97. Background and quote from: Rodrigues Fowler Y. and Goodman C., 'How Tinder Could Take Back the White House', *New York Times*, 22 June 2017.

98. Solon O. and Siddiqui S., 'Russia-backed Facebook posts "reached 126m Americans" during US election', *The Guardian*, 31 October

2017; Statt N., 'Twitter says it exposed nearly 700,000 people to Russian propaganda during US election', *The Verge*, 19 January 2018.

99. Watts D.J. and Rothschild D.M., 'Don't blame the election on fake news. Blame it on the media', *Columbia Journalism Review*, 2017. See also: Persily N. and Stamos A., 'Regulating Online Political Advertising by Foreign Governments and Nationals', in McFaul M. (ed.), 'Securing American Elections', Stanford University, June 2019.

100. Confessore N. and Yourish K., '$2 Billion Worth of Free Media for Donald Trump', *New York Times*, 16 March 2016.

101. Sources: Guess A. et al., 'Selective Exposure to Misinformation: Evidence from the consumption of fake news during the 2016 U.S. presidential campaign', 2018; Guess A. et al., 'Fake news, Facebook ads, and misperceptions: Assessing information quality in the 2018 U.S. midterm election campaign', 2019; Narayanan V. et al., 'Russian Involvement and Junk News during Brexit', *Oxford Comprop Data Memo*, 2017.

102. Pareene A., 'How We Fooled Donald Trump Into Retweeting Benito Mussolini', *Gawker*, 28 February 2016.

103. Hessdec A., 'On Twitter, a Battle Among Political Bots', *New York Times*, 14 December 2016.

104. Shao C. et al., 'The spread of low-credibility content by social bots', *Nature Communications*, 2018.

105. Musgrave S., 'ABC, AP and others ran with false information on shooter's ties to extremist groups', *Politico*, 16 February 2018.

106. O'Sullivan D., 'American media keeps falling for Russian trolls', *CNN*, 21 June 2018.

107. Phillips W., 'How journalists should not cover an online conspiracy theory', *The Guardian*, 6 August 2018.

108. Background on media manipulation from: Phillips W., 'The Oxygen of Amplification', *Data & Society Report*, 2018.

109. Weiss M., 'Revealed: The Secret KGB Manual for Recruiting Spies', *The Daily Beast*, 27 December 2017.

110. DiResta R., 'There are bots. Look around', *Ribbon Farm*, 23 May 2017.

111. 'Over 9000 Penises', *Know Your Meme*, 2008.

112. Zannettou S. et al., 'On the Origins of Memes by Means of Fringe Web Communities', *arXiv*, 2018.

113. Feinberg A., 'This is the Daily Stormer's playbook', *Huffington Post*, 13 December 2017.

114. Collins K. and Roose K., 'Tracing a Meme From the Internet's Fringe to a Republican Slogan', *New York Times*, 4 November 2018.

115. Background on real-life spillover: O'Sullivan D., 'Russian trolls created Facebook events seen by more than 300,000 users', *CNN*, 26 January 2018; Taub A. and Fisher M., 'Where Countries Are Tinderboxes and Facebook Is a Match', *New York Times*, 21 April 2018. Analysis of the #BlackLivesMatter online movement also uncovered Russian accounts contributing to both sides of the debate: Stewart L.G. et al., 'Examining Trolls and Polarization with a Retweet Network', *MIS2*, 2018.

116. Broniatowski D.A. et al., 'Weaponized Health Communication: Twitter Bots and Russian Trolls Amplify the Vaccine Debate', *American Journal of Public Health*, 2018; Wellcome Global Monitor 2018, 19 June 2019.

117. Google Ngram.

118. Takayasu M. et al., 'Rumor Diffusion and Convergence during the 3.11 Earthquake: A Twitter Case Study', *PLOS ONE*, 2015.

119. Friggeri A. et al., 'Rumor Cascades'. *AAAI Publications*, 2014.

120. 'WhatsApp suggests a cure for virality', *The Economist*, 26 July 2018.

121. McMillan R. and Hernandez D., 'Pinterest Blocks Vaccination Searches in Move to Control the Conversation', *Wall Street Journal*, 20 February 2019.

122. Quotes from author interview with Whitney Phillips, October 2018.

123. Baumgartner J. et al., 'What we learned from analyzing thousands of stories on the Christchurch shooting', *Columbia Journalism Review*, 2019.

124. Quotes from author interview with Brendan Nyhan, November 2018.

125. Source: Web of Science. Search string: (<platform> AND (contagio* OR diffus* OR transmi*). Studies were excluded if they only mentioned the platform as an illustrative or comparative

example, or focused adoption of the platform itself rather than diffusion via the platform. In total, 391 Twitter studies and 85 Facebook studies during 2016–2018. 330m Twitter users in 2019 vs 2,400m Facebook users. Source for user data: https://www.statista. com/

126. Nelson A. et al., 'The Social Science Research Council Announces the First Recipients of the Social Media and Democracy Research Grants', *Social Sciences Research Council Items*, 29 April 2019; Alba D., 'Ahead of 2020, Facebook Falls Short on Plan to Share Data on Disinformation', *New York Times*, 29 September 2019.

127. 'Almost all of Vote Leave's digital communication and data science was invisible even if you read every single news story or column ever produced in the campaign or any of the books so far published'. Quote from: Cummings D., 'On the referendum #20', Dominic Cummings's Blog, 29 October 2016. In October 2018, Facebook established a public archive of political adverts – an important shift, although it still only captures the first step of the information transmission processes. Source: Cellan-Jones R., 'Facebook tool makes UK political ads "transparent"', BBC News Online, 16 October 2018.

128. Ginsberg D. and Burke M., 'Hard Questions: Is Spending Time on Social Media Bad for Us?' Facebook newsroom, 15 December 2017; Burke M. et al., 'Social Network Activity and Social Well-Being', *Proceedings of the 28th International Conference on Human Factors in Computing Systems*, 2010; Burke M. and Kraut R.E., 'The Relationship Between Facebook Use and Well-Being Depends on Communication Type and Tie Strength', *Journal of Computer-Mediated Communication*, 2016.

129. Routledge I. et al., 'Estimating spatiotemporally varying malaria reproduction numbers in a near elimination setting', *Nature Communications*, 2018.

6. How to own the internet

1. Background on Mirai from: Antonakakis M. et al., 'Understanding the Mirai Botnet', *Proceedings of the 26th USENIX Security Symposium*, 2017; Solomon B. and Fox-Brewster T., 'Hacked Cameras Were Behind Friday's Massive Web Outage', *Forbes*, 21

October 2016; Bours B., 'How a Dorm Room Minecraft Scam Brought Down the Internet', *Wired*, 13 December 2017.

2. Quoted in: Bours B., 'How a Dorm Room Minecraft Scam Brought Down the Internet', *Wired*, 13 December 2017.

3. Background on WannaCry from: 'What you need to know about the WannaCry Ransomware', *Symantec Blogs*, 23 October 2017; Field M., 'WannaCry cyber attack cost the NHS £92m as 19,000 appointments cancelled', *The Telegraph*, 11 October 2018; Wiedeman R., 'The British hacker Marcus Hutchins and the FBI', *The Times*, 7 April 2018.

4. Moore D. et al., 'The Spread of the Sapphire/Slammer Worm', *Center for Applied Internet Data Analysis* (CAIDA), 2003.

5. Background on Elk Cloner from: Leyden J., 'The 30-year-old prank that became the first computer virus', *The Register*, 14 December 2012.

6. Quotes from author interview with Alex Vespignani, May 2018.

7. Cohen F., 'Computer Viruses – Theory and Experiments', 1984.

8. Background on Morris Worm from: Seltzer L., 'The Morris Worm: Internet malware turns 25', *Zero Day*, 2 November 2013; UNITED STATES of America, Appellee, v. Robert Tappan MORRIS, Defendant-appellant. 928 F.2D 504, 1990.

9. Graham P., 'The Submarine', April 2005. http://www.paulgraham.com

10. Moon M., '"Minecraft" success helps its creator buy a $70 million mansion', *Engadget*, 18 December 2014.

11. Background on DDoS from: 'Who is Anna-Senpai, the Mirai Worm Author?', *Krebs on Security*, 18 January 2017; 'Spreading the DDoS Disease and Selling the Cure', 19 October 2016.

12. 'Computer Hacker Who Launched Attacks On Rutgers University Ordered To Pay $8.6m', U.S. Attorney's Office, District of New Jersey, 26 October 2018.

13. @MalwareTechBlog, 13 May 2017.

14. Staniford S. et al., 'How to own the Internet in Your Spare Time', *ICIR*, 2002.

15. Assuming R=20 and infectious for 8 days, equivalent to 0.1 infections per hour.

16. Moore D. et al., 'The Spread of the Sapphire/Slammer Worm', *Center for Applied Internet Data Analysis* (CAIDA), 2003.

17. 'Kaspersky Lab Research Reveals the Cost and Profitability of Arranging a DDoS Attack', Kaspersky Lab, 23 March 2017.

18. Palmer D., 'Ransomware is now big business on the dark web and malware developers are cashing in', *ZDNet*, 11 October 2017.

19. Nakashima E. and Timberg C., 'NSA officials worried about the day its potent hacking tool would get loose. Then it did', *Washington Post*, 16 May 2017.

20. Orr A., 'Zerodium Offers $2 Million for Remote iOS Exploits', *Mac Observer*, 10 January 2019.

21. Background on Student from: Kushner D., 'The Real Story of Stuxnet', *IEEE Spectrum*, 26 February 2013; Kopfstein J., 'Stuxnet virus was planted by Israeli agents using USB sticks, according to new report', *The Verge*, 12 April 2012.

22. Kaplan F., *Dark Territory: The Secret History of Cyber War* (Simon & Schuster, 2016).

23. Dark Trace. Global Threat Report 2017. http://www.darktrace.com

24. Background and quotes from: Lomas A., 'Screwdriving. Locating and exploiting smart adult toys', *Pen Test Partners Blog*, 29 September 2017; Franceschi-Bicchierai L., 'Hackers Can Easily Hijack This Dildo Camera and Livestream the Inside of Your Vagina (Or Butt)', *Motherboard*, 3 April 2017.

25. DeMarinis N. et al., 'Scanning the Internet for ROS: A View of Security in Robotics Research', *arXiv*, 2018.

26. Background on AWS outage from: Hindi R., 'Thanks for breaking our connected homes, Amazon', *Medium*, 28 February, 2017; Hern A., 'How did an Amazon glitch leave people literally in the dark?', *The Guardian*, 1 March 2017.

27. Background on AWS performance: Amazon Compute Service Level Agreement. https://aws.amazon.com, 12 February 2018; Poletti T., 'The engine for Amazon earnings growth has nothing to do with e-commerce', *Market Watch*, 29 April 2018.

28. Swift D., '"Mega Outage" Wreaks Havoc on Internet, is AWS too

Big to Fail?', *Digit*, 2017; Bobeldijk Y., 'Is Amazon's cloud service too big to fail?', *Financial News*, 1 August 2017.

29. Barrett B. and Newman L.H., 'The Facebook Security Meltdown Exposes Way More Sites Than Facebook', *Wired*, 28 September 2018.

30. Background on Love Bug: Meek J., 'Love bug virus creates worldwide chaos', *The Guardian*, 5 May 2000; Barabási A.L., *Linked: the New Science of Networks* (Perseus Books, 2003).

31. White S.R., 'Open Problems in Computer Virus Research', *Virus Bulletin Conference*, 1998.

32. Barabási A.L. and Albert R., 'Emergence of Scaling in Random Networks', *Science*, 1999.

33. Pastor-Satorras R. and Vespignani A., 'Epidemic Spreading in Scale-Free Networks', *Physical Review Letters*, 2 April 2001.

34. Goel S. et al., 'The Structural Virality of Online Diffusion', *Management Science*, 2016.

35. Background on left-pad from: Williams C., 'How one developer just broke Node, Babel and thousands of projects in 11 lines of JavaScript', *The Register*, 23 March 2016; Tung L., 'A row that led a developer to delete a 17-line JavaScript module has stopped countless applications working', *ZDNet*, 23 March 2016; Roberts M., 'A discussion about the breaking of the Internet', *Medium*, 23 March 2016.

36. Haney D., 'NPM & left-pad: Have We Forgotten How To Program?' 23 March 2016, https://www.davidhaney.io

37. Rotabi R. et al., 'Tracing the Use of Practices through Networks of Collaboration', *AAAI*, 2017.

38. Fox-Brewster T., 'Hackers Sell $7,500 IoT Cannon To Bring Down The Web Again', *Forbes*, 23 October 2016.

39. Gallagher S., 'New variants of Mirai botnet detected, targeting more IoT devices', *Ars Technica*, 9 April 2019.

40. Cohen F., 'Computer Viruses – Theory and Experiments', 1984.

41. Cloonan J., 'Advanced Malware Detection – Signatures vs. Behavior Analysis', *Infosecurity Magazine*, 11 April 2017.

42. Oldstone M.B.A., *Viruses, Plagues, and History* (Oxford University Press, 2010).

43. Background on Beebone from: Goodin D., 'US, European police take down highly elusive botnet known as Beebone', *Ars Technica*, 9 April 2015; Samani R., 'Update on the Beebone Botnet Takedown', *McAfee Blogs*, 20 April 2015.

44. Thompson C.P. et al., 'A naturally protective epitope of limited variability as an influenza vaccine target', *Nature Communications*, 2018.

45. 'McAfee Labs 2019 Threats Predictions Report', McAfee Labs, 29 November 2018; Seymour J. and Tully P., 'Weaponizing data science for social engineering: Automated E2E spear phishing on Twitter', Working paper, 2016.

7. Tracking outbreaks

1. Background on Schmidt case from: Court of Appeal of Louisiana, Third Circuit. STATE of Louisiana v. Richard J. SCHMIDT. No. 99–1412, 2000; Miller M., 'A Deadly Attraction', *Newsweek*, 18 August 1996.

2. Darwin C., *Journal of researches into the natural history and geology of the countries visited during the voyage of H.M.S. Beagle round the world, under the command of Capt. Fitz Roy, R.N.* (John Murray, 1860).

3. Hon C.C. et al., 'Evidence of the Recombinant Origin of a Bat Severe Acute Respiratory Syndrome (SARS)-Like Coronavirus and Its Implications on the Direct Ancestor of SARS Coronavirus', *Journal of Virology*, 2008.

4. Forensic File Update on Janice Trahan Case, CNN, 14 March 2016.

5. González-Candelas F. et al., 'Molecular evolution in court: analysis of a large hepatitis C virus outbreak from an evolving source', *BMC Biology*, 2013; Fuchs D., 'Virus doctor jailed for 1,933 years', *The Guardian*, 16 May 2007.

6. Oliveira T. et al., 'HIV-1 and HCV sequences from Libyan outbreak', *Nature*, 2006; 'HIV medics released to Bulgaria', BBC News Online, 24 July 2007.

7. Köser C.U. et al., 'Rapid Whole-Genome Sequencing for Investigation of a Neonatal MRSA Outbreak', *NEJM*, 2012; Fraser C. et al., 'Pandemic Potential of a Strain of Influenza A (H1N1): Early Findings', *Science*, 2009.

8. Kama M. et al., 'Sustained low-level transmission of Zika and

chikungunya viruses following emergence in the Fiji Islands, Pacific', *Emerging Infectious Diseases*, 2019.

9. Diallo B. et al., 'Resurgence of Ebola virus disease in Guinea linked to a survivor with virus persistence in seminal fluid for more than 500 days', *Clinical Infectious Diseases*, 2016.

10. Racaniello V., 'Zika virus, like all other viruses, is mutating', *Virology Blog*, 14 April 2016.

11. Beaty B.M. and Lee B., 'Constraints on the Genetic and Antigenic Variability of Measles Virus', *Viruses*, 2016.

12. Background on sequence availability: Gire S.K. et al., 'Genomic surveillance elucidates Ebola virus origin and transmission during the 2014 outbreak', *Science*, 2014; Yozwiak N.L., 'Data sharing: Make outbreak research open access', *Nature*, 2015; Gytis Dudas, https://twitter.com/evogytis/status/1065157012261126145

13. Sample I., 'Thousands of lives put at risk by clinical trials system that is "not fit for purpose"', *The Guardian*, 31 March 2014.

14. Callaway E., 'Zika-microcephaly paper sparks data-sharing confusion', *Nature*, 12 February 2016; Maxmen, A., 'Two Ebola drugs show promise amid ongoing outbreak,' *Nature*, 12 August 2019; Johansson M.A. et al., 'Preprints: An underutilized mechanism to accelerate outbreak science', *PLOS Medicine*, 2018; https://nextstrain.org/community/inrb-drc/ebola-nord-kivu

15. Sabeti P., 'How we'll fight the next deadly virus', *TEDWomen* 2015.

16. Hadfield J. et al., 'Nextstrain: real-time tracking of pathogen evolution', *Bioinformatics*, 2018.

17. Doughton S., '250,000 people now follow this Fred Hutch scientist on Twitter. We talk to this leading voice of the coronavirus pandemic', *Seattle Times*, 1 June 2020.

18. https://twitter.com/trvrb/status/1233970271318503426?s=20

19. Owlcation, 'The History Behind the Story of Goldilocks', 22 February 2018, https://owlcation.com/humanities/goldilocks-and-three-bears

20. Background and quotes from author interview with Jamie Tehrani, October 2017.

21. Tehrani J.J., 'The Phylogeny of Little Red Riding Hood', *PLOS ONE*, 2013.

22. Van Wyhe J., 'The descent of words: evolutionary thinking 1780–1880', *Endeavour*, 2005.

23. Luu C., 'The Fairytale Language of the Brothers Grimm', *JSTOR Daily*, 2 May 2018.

24. Da Silva S.G. and Tehrani J.J., 'Comparative phylogenetic analyses uncover the ancient roots of Indo-European folktales', *Royal Society Open Science*, 2015.

25. Smith D. et al., 'Cooperation and the evolution of hunter-gatherer storytelling', *Nature Communications*, 2017.

26. Background from: Stubbersfield J.M. et al., 'Serial killers, spiders and cybersex: social and survival information bias in the transmission of urban legends', *British Journal of Psychology*, 2015. A similar result pattern has been found in other telephone studies, with social information seemingly having an advantage when it comes to transmission.

27. Background on counter-intuitive elements from: Mesoudi A. and Whiten A., 'The multiple roles of cultural transmission experiments in understanding human cultural evolution', *Philosphical Transactions of the Royal Society B*, 2008; Stubbersfield J. and Tehrani J., 'Expect the Unexpected? Testing for Minimally Counterintuitive (MCI) Bias in the Transmission of Contemporary Legends: A Computational Phylogenetic Approach', *Social Science Computer Review*, 2013.

28. Dlugan A., 'How to Use the Rule of Three in Your Speeches', 27 May 2009. http://sixminutes.dlugan.com/rule-of-three-speeches-public-speaking

29. The rule of three is also common in comedy, where an unexpected third item creates the punchline.

30. Newberry M.G. et al., 'Detecting evolutionary forces in language change', *Nature*, 2017.

31. Valverde S. and Sole R.V., 'Punctuated equilibrium in the large-scale evolution of programming languages', *Journal of the Royal Society Interface*, 2015.

32. Svinti V. et al., 'New approaches for unravelling reassortment pathways', *BMC Evolutionary Biology*, 2013.

33. Sample I., 'Evolution: Charles Darwin was wrong about the tree of life', *The Guardian*, 21 January 2009.

34. Background on sponging from: Krützen M. et al., 'Cultural transmission of tool use in bottlenose dolphins', *PNAS*, 2005; Morell V., 'Why Dolphins Wear Sponges', *Science*, 20 July 2011.

35. Background and quotes from author interview with Lucy Aplin, August 2017.

36. Baker K.S. et al., 'Horizontal antimicrobial resistance transfer drives epidemics of multiple Shigella species', *Nature Communications*, 2018; McCarthy A.J. et al., 'Extensive Horizontal Gene Transfer during Staphylococcus aureus Co-colonization In Vivo', *Genome Biology and Evolution*, 2014; Alirol E. et al., 'Multidrug-resistant gonorrhea: A research and development roadmap to discover new medicines', *PLOS Medicine*, 2017.

37. Gallagher J., 'Man has "world's worst" super-gonorrhoea', BBC News Online, 28 March 2018; Gallagher J., 'Super-gonorrhoea spread causes "deep concern"', BBC News Online, 9 January 2019.

38. Alzheimer's Society's view on genetic testing. April 2015. https://www.alzheimers.org.uk/about-us/policy-and-influencing/what-we-think/genetic-testing; Genetic testing for cancer risk. Cancer Research UK. https://www.cancerresearchuk.org/about-cancer/causes-of-cancer/inherited-cancer-genes-and-increased-cancer-risk/genetic-testing-for-cancer-risk

39. Middleton A., 'Attention The Times: Prince William's DNA is not a toy', *The Conversation*, 14 June 2013. Researchers have also criticised the scientific analysis behind the story. Source: Kennett D.A, 'The Rise and Fall of Britain's DNA: A Tale of Misleading Claims, Media Manipulation and Threats to Academic Freedom', *Genealogy*, 2018.

40. Ash L., 'The Christmas present that could tear your family apart', BBC News Online, 20 December 2018.

41. Clark K., 'Scoop: 23andMe is raising up to $300M', *PitchBook*, 24 July 2018; Rutherford A., 'DNA ancestry tests may look cheap. But your data is the price', *The Guardian*, 10 August 2018.

42. Cox N., 'UK Biobank shares the promise of big data', *Nature*, 10 October 2018.

43. Based on 1990 census data, Sweeney estimated 87 per cent of

people could be identified. Subsequent studies revised this down to 61–63 per cent based on 1990 and 2000 data. Background: Sweeney L., 'Simple Demographics Often Identify People Uniquely', Carnegie Mellon University, Data Privacy Working Paper, 2000; Ohm P., 'Broken Promises of Privacy: Responding to the Surprising Failure of Anonymization', *UCLA Law Review*, 2010; Sweeney L., 'Only You, Your Doctor, and Many Others May Know', *Technology Science*, 2015.

44. Sweeney L., 'Only You, Your Doctor, and Many Others May Know', *Technology Science*, 2015.

45. Smith S., 'Data and privacy', *Significance*, 3 October 2014.

46. Background on taxi data from: Whong C., 'FOILing NYC's Taxi Trip Data', 18 March 2014. https://chriswhong.com; Pandurangan V., 'On Taxis and Rainbows', 21 June 2014. https://tech.vijayp.ca

47. Background and quotes from: Tockar A., 'Riding with the Stars: Passenger Privacy in the NYC Taxicab Dataset', 15 September 2014. https://research.neustar.biz.

48. De Montjoye Y.A., 'Unique in the Crowd: The privacy bounds of human mobility', *Scientific Reports*, 2013.

49. Shahani A., 'Smartphones Are Used To Stalk, Control Domestic Abuse Victims', National Public Radio, 15 September 2014.

50. Hern A., 'Fitness tracking app Strava gives away location of secret US army bases', *The Guardian*, 28 January 2014.

51. Watts A.G. et al., 'Potential Zika virus spread within and beyond India', *Journal of Travel Medicine*, 2018; Bengtsson L. et al., 'Improved Response to Disasters and Outbreaks by Tracking Population Movements with Mobile Phone Network Data: A Post-Earthquake Geospatial Study in Haiti', *PLOS Medicine*, 2011; Santi P. et al., 'Quantifying the benefits of vehicle pooling with shareability networks', *PNAS*, 2014.

52. Chen M.K. and Rohla R., 'The effect of partisanship and political advertising on close family ties', *Science*, 2018; Silm S. et al., 'Are younger age groups less segregated? Measuring ethnic segregation in activity spaces using mobile phone data', *Journal of Ethnic and Migration Studies*, 2017; Xiao Y. et al., 'Exploring the disparities in park access through mobile phone data: Evidence from Shanghai,

China', *Landscape and Urban Planning*, 2019; Atlas of Inequality, https://inequality.media.mit.edu.

53. Conlan A.J.K. et al., 'Measuring social networks in British primary schools through scientific engagement', *Proceedings of the Royal Society B*, 2010.

54. Background on GPS brokers from: Harris R., 'Your Apps Know Where You Were Last Night, and They're Not Keeping It Secret', *New York Times*, 10 December 2018; Signoret P., Teemo, 'la start-up qui traque 10 millions de Français en continu', *L'Express L'Expansion*, 25 August 2018; 'Is Geospatial Data a $100 Billion Business for SafeGraph?' *Nanalyze*, 22 April 2017.

55. Importantly, the target gave permission for their phone to be tracked. Source: Cox J., 'I Gave a Bounty Hunter $300. Then He Located Our Phone', *Motherboard*, 8 January 2019.

56. Scam alert: Speeding ticket email scam. Tredyffrin Police Department. 23 March 2016.

57. Background on SARS introduction from: 'SARS Commission Final Report', Government of Ontario, 2005; Tsang K.W. et al., 'A Cluster of Cases of Severe Acute Respiratory Syndrome in Hong Kong', *The NEJM*, 2003.

58. Donnelly C.A. et al., 'Epidemiological determinants of spread of causal agent of severe acute respiratory syndrome in Hong Kong', *The Lancet*, 2003.

59. WHO Ebola Response Team, 'Ebola Virus Disease in West Africa – The First 9 Months of the Epidemic and Forward Projections', *NEJM*, 2014; Assiri A. et al., 'Hospital Outbreak of Middle East Respiratory Syndrome Coronavirus', *NEJM*, 2013; WHO Consultation on Clinical Aspects of Pandemic (H1N1) 2009 Influenza, 'Clinical Aspects of Pandemic 2009 Influenza A (H1N1) Virus Infection', *NEJM*, 2010; Lauer SA et al., The Incubation Period of Coronavirus Disease 2019 (COVID-19) From Publicly Reported Confirmed Cases: Estimation and Application. Annals of Internal Medicine https://annals.org/aim/fullarticle/2762808/incubation-period-coronavirus-disease-2019-covid-19-from-publicly-reported.

60. Background on Willowbrook from: Rothman D.J., *The Willowbrook Wars: Bringing the Mentally Disabled into the Community* (Aldine

Transaction, 2005); Fansiwala K., 'The Duality of Medicine: The Willowbrook State School Experiments', *Medical Dialogue Review*, 20 February 2016; Watts G., 'Robert Wayne McCollum', *The Lancet*, 2010.

61. Quoted in: Offit P., *Vaccinated: One Man's Quest to Defeat the World's Deadliest Diseases* (Harper Perennial, 2008).

62. Goldby S., 'Experiments at the Willowbrook state school', *The Lancet*, 1971.

63. Gordon R.M., *The Infamous Burke and Hare: Serial Killers and Resurrectionists of Nineteenth Century Edinburgh* (McFarland, 2009).

64. Transcript for NMT 1: Medical Case, 9 January 1947. Harvard Law School Library Nuremberg Trials Project.

65. Waddington C.S. et al., 'Advancing the management and control of typhoid fever: A review of the historical role of human challenge studies', *Journal of Infection*, 2014.

66. Background on modern challenge studies: Cohen J., 'Studies that intentionally infect people with disease-causing bugs are on the rise', *Science*, 18 May 2016; https://clinicaltrials.gov; Nordling L., 'The Ethical Quandary of Human Infection Studies', *Undark*, 19 November 2018.

8. A spot of trouble

1. Peterson Hill N., *A Very Private Public Citizen: The Life of Grenville Clark* (University of Missouri, 2016).

2. Ham P., 'As Hiroshima Smouldered, Our Atom Bomb Scientists Suffered Remorse', *Newsweek*, 5 August 2015.

3. Ito S., 'Einstein's pacifist dilemma revealed', *The Guardian*, 5 July 2005; 'The Einstein Letter That Started It All; A message to President Roosevelt 25 Years ago launched the atom bomb and the Atomic Age', *New York Times*, 2 August 1964.

4. Clark G., Letters to the Times, *New York Times*, 22 April 1955.

5. Harris E.D. et al., 'Governance of Dual-Use Technologies: Theory and Practice', *American Academy of Arts & Sciences*, 2016.

6. Santi P. et al., 'Quantifying the benefits of vehicle pooling with shareability networks', *PNAS*, 2014; other references covered in earlier chapters.

7. Cadwalladr C. et al., 'Revealed: 50 million Facebook profiles

harvested for Cambridge Analytica in major data breach', *The Guardian*, 17 March 2018.

8. Sumpter S., *Outnumbered: From Facebook and Google to Fake News and Filter-bubbles – The Algorithms That Control Our Lives* (Bloomsbury Sigma, 2018); Chen A. et al., 'Cambridge Analytica's Facebook data abuse shouldn't get credit for Trump', *The Verge*, 20 March 2018.

9. Zunger Y., 'Computer science faces an ethics crisis. The Cambridge Analytica scandal proves it', *Boston Globe*, 22 March 2018.

10. Harkin J., '"Big Data", "Who Owns the Future?" and "To Save Everything, Click Here"', *Financial Times*, 1 March 2013; Harford T., 'Big data: A big mistake?', *Significance*, 1 December 2014; McAfee A. et al., 'Big Data: The Management Revolution', *Harvard Business Review*, October 2012.

11. Ginsberg J. et al., 'Detecting influenza epidemics using search engine query data', *Nature*, 2009.

12. Olson D.R. et al., 'Reassessing Google Flu Trends Data for Detection of Seasonal and Pandemic Influenza: A Comparative Epidemiological Study at Three Geographic Scales', *PLOS Computational Biology*, 2013.

13. Lazer D. et al., 'The Parable of Google Flu: Traps in Big Data Analysis,' *Science*, 2014.

14. World Health Organization, 'Pandemic influenza vaccine manufacturing process and timeline', *WHO Briefing Note*, 2009.

15. Petrova V.N. et al., 'The evolution of seasonal influenza viruses', *Nature Reviews Microbiology*, 2017; Chakraborty P. et al., 'What to know before forecasting the flu', *PLOS Computational Biology*, 2018.

16. Buckee C., 'Sorry, Silicon Valley, but "disruption" isn't a cure-all', *Boston Globe*, 22 January 2017.

17. Farrar J., 'The key to fighting the next "Ebola" outbreak is in your pocket', *Wired*, 4 December 2016; other references covered in earlier chapters.

18. World Health Organisation, 'Ebola outbreak in the Democratic Republic of the Congo declared a Public Health Emergency of International Concern', WHO newsroom, 17 July 2019; Silberner J., 'Congo's fight against Ebola stalls after epidemiologist is shot dead', *British Medical Journal*, 2019.

19. Ginsberg M. et al., 'Swine Influenza A (H1N1) Infection in Two Children – Southern California, March–April 2009, *Morbidity and Mortality Weekly Report*, 2009.

20. Cohen J., 'As massive Zika vaccine trial struggles, researchers revive plan to intentionally infect humans', *Science*, 12 September 2018; Koopmans M. et al., 'Familiar barriers still unresolved – a perspective on the Zika virus outbreak research response', *The Lancet Infectious Diseases*, 2018.

21. Gordon A. et al., 'Prior dengue virus infection and risk of Zika: A pediatric cohort in Nicaragua', *PLOS Medicine*, 2019.

22. Grinberg N. et al., 'Fake news on Twitter during the 2016 U.S. presidential election', *Science*, 2019; Guess A. et al., 'Less than you think: Prevalence and predictors of fake news dissemination on Facebook', *Science Advances*, 2019; Lazer D.M.J. et al., 'The science of fake news', *Science*, 2018; Wagner K., 'Inside Twitter's ambitious plan to change the way we tweet', *Recode*, 8 March 2019; McCarthy K., 'Facebook, Twitter slammed for deleting evidence of Russia's US election mischief', *The Register*, 13 October 2017.

23. Haldane A.G., 'Rethinking the Financial Network', Bank of England speech, 28 April 2009; Editorial Board, 'A fractured reporting system stymies public-safety research', *Bloomberg*, 25 October 2018.

24. Greene G., *The Woman Who Knew Too Much: Alice Stewart and the Secrets of Radiation* (University of Michigan Press, 2001).

25. Presentation at Epidemics[6] conference, 2017.

26. Kosinski M. et al., 'Private traits and attributes are predictable from digital records of human behavior', *PNAS*, 2013.

27. Cadwalladr C. et al., 'Revealed: 50 million Facebook profiles harvested for Cambridge Analytica in major data breach', *The Guardian*, 17 March 2018. Note that despite the apparent similarity in methods, Cambridge Analytica did not work with Kosinki.

28. Alaimo K., 'Twitter's Misguided Barriers for Researchers', *Bloomberg*, 16 October 2018.

29. Godlee F., 'What can we salvage from care.data?', *British Medical Journal*, 2016.

30. Kucharski A.J. et al., 'School's out: seasonal variation in the

movement patterns of school children', *PLOS ONE*, 2015; Kucharski A.J. et al., 'Structure and consistency of self-reported social contact networks in British secondary schools', *PLOS ONE*, 2018.

31. http://www.bbc.co.uk/pandemic.

32. Information Commissioner's Office, 'Investigation into the use of data analytics in political campaigns', *ICO report*, 11 July 2018.

33. Rifkind H., TV review, *The Times*, 24 March 2018.

34. BBC News Online. 'Coronavirus: Latest patient was first to be infected in UK', 29 February 2020.

35. Kucharski A. J. et al., 'Effectiveness of isolation, testing, contact tracing and physical distancing on reducing transmission of SARS-CoV-2 in different settings: a mathematical modelling study', *Lancet Inf Dis*, 2020; Jarvis C. I. et al., 'Quantifying the impact of physical distance measures on the transmission of COVID-19 in the UK', *BMC Medicine*, 2020.

36. Assuming a six-hour reading time (i.e. 225 words per minute). Data: World Health Organization. http://www.who.int, 2018; Dance D.A. et al., 'Global Burden and Challenges of Melioidosis', *Tropical Medicine and Infectious Disease*, 2018.

37. Declined from 291 per 100,000 in 1990 to 154 per 100,000 in 2016. Source: Ritchie H. et al., 'Causes of Death', *Our World in Data*, 2018.

38. UK Government, *Health profile for England: 2017*. https://www.gov.uk.

39. Harper-Jemison D.M. et al., 'Leading causes of death in Chicago', Chicago Department of Public Health Office of Epidemiology, 2006; 'Illinois State Fact Sheet', National Injury and Violence Prevention Resource Center, 2015.

40. Information Commissioner's Office, 'Investigation into the use of data analytics in political campaigns', *ICO report*, 11 July 2018; DiResta R. et al., 'The Tactics & Tropes of the Internet Research Agency', *New Knowledge*, 2018.

Further reading

If you want to know more about the topics covered in this book, additional suggestions for articles, papers and books are given below. To ensure reproducibility, all data and code required to generate the figures in the book are available from: https://github.com/adamkucharski/rules-of-contagion/

Chapter 1

A trio of papers by Paul Fine has more information on the theory of mechanistic modelling and resulting concepts like herd immunity: 'Ross's A Priori Pathometry – A Perspective' (*Proceedings of the Royal Society of Medicine*, 1975); 'John Brownlee and the measurement of infectiousness: an historical study in epidemic theory' (*Journal of the Royal Statistical Society: Series A*, 1979); 'Herd Immunity: History, Theory, Practice' (*Epidemiological Reviews*, 1993). For a more technical description of Ross's analysis and its legacy, see David Smith and colleagues' paper 'Ross, Macdonald, and a Theory for the Dynamics and Control of Mosquito-Transmitted Pathogens' (*PLOS Pathogens*, 2012).

Chapter 2

Donald MacKenzie and Taylor Spears's paper '"The Formula That Killed Wall Street"?: The Gaussian Copula and the Material Cultures of Modelling' (2012) provides a useful oral history of the models behind CDOs. *Liar's Poker: Rising Through the Wreckage on Wall Street* (W. W. Norton & Company, 1989) and *The Big Short: Inside the Doomsday Machine* (W. W.

Norton & Company, 2010) by Michael Lewis, written twenty years apart, explain how mortgage trading started and the chaos it would later cause. *When Genius Failed: The Rise and Fall of Long-Term Capital Management* by Roger Lowenstein (Random House, 2000) covers the collapse of the titular hedge fund.

Seeking the Positives: A Life Spent on the Cutting Edge of Public Health by John Potterat (CreateSpace, 2015) gives more details of his work on how social networks shape outbreaks of gonorrhoea and other STDs. For a technical overview of disease modelling, *Modelling Infectious Diseases in Humans and Animals* (Princeton University Press, 2007) by Matt Keeling and Pej Rohani had been an essential textbook for me ever since I first read it as an undergraduate.

Andy Haldane's speech 'Rethinking the Financial Network' (Bank of England transcript, 2009) was a timely discussion of the links between ecology, epidemiology and financial markets. His later paper with Robert May, 'Systemic risk in banking ecosystems' (*Nature*, 2011), expanded on these ideas with more technical details.

Chapter 3

Connected: The Amazing Power of Social Networks and How They Shape Our Lives by Nicholas Christakis and James Fowler (HarperPress, 2011) describes research into dynamics of social networks, including their studies on the spread of obesity and other characteristics. Their subsequent paper 'Social contagion theory: examining dynamic social networks and human behavior' (*Statistics in Medicine*, 2013) discusses the criticisms of their research, and the technical challenges involved in estimating social contagion. Damon Centola's book *How Behavior Spreads: The Science of Complex Contagions* (Princeton University Press, 2018) covers his work on complex contagion, as well as other insights from large-scale studies of behaviour. 'Randomized experiments to detect and estimate social influence in networks' by Sean Taylor and Dean Eckles (*Complex Spreading Phenomena in Social Systems*, 2018) is a useful technical review of approaches for studying social contagion.

Further insights from the NATSAL studies can be found in David Spiegelhalter's book *Sex by Numbers: What Statistics Can Tell Us About Sexual Behaviour* (Wellcome Collection, 2015). 'Culture and cultural evolution in birds: a review of the evidence' by Lucy Aplin (*Animal Behaviour*,

2019) provides an overview of cultural development in animals, with a focus on birds.

Chapter 4

For more discussion and case studies about the spread of violence, including contributions from Carl Bell, Gary Slutkin and Charlotte Watts, see the papers published in *Contagion of Violence: Workshop Summary*, part of the Forum on Global Violence Prevention (The National Academies Collection, 2013).

Smallpox: The Death of a Disease – The Inside Story of Eradicating a Worldwide Killer by D.A. Henderson (Prometheus, 2009) has a first-hand account of how contact tracing and ring vaccination was deployed to eradicate smallpox. Neil Ferguson and colleagues' paper 'Planning for smallpox outbreaks' (*Nature*, 2003) covers ways to model smallpox and other emerging infections, as well as their limitations. 'Avoidable errors in the modelling of outbreaks of emerging pathogens, with special reference to Ebola' by Aaron King and colleagues (*Proceedings of the Royal Society B*, 2015) provides a technical description of some potential pitfalls in forecasting infectious disease outbreaks.

Weapons of Math Destruction: How Big Data Increases Inequality and Threatens Democracy by Cathy O'Neil (Penguin, 2016) highlights the inherent prejudices and biases in many commonly used algorithms, including ones used in policing. *Hello World: How to be Human in the Age of the Machine* by Hannah Fry (Penguin, 2019) has more on the roles – and risks – of algorithms in modern life.

Chapter 5

Duncan Watts' book *Everything is Obvious: Why Common Sense is Nonsense* (Atlantic Books, 2011) has some useful insights into the challenges of understanding and predicting social behaviour online. His later paper with Jake Hofman and Amit Sharma, 'Prediction and explanation in social systems' (*Science*, 2017), elaborates on the technical aspects of this research. Justin Cheng and colleagues' paper 'Do Diffusion Protocols Govern Cascade Growth?' (AAAI, 2018) provides a data-driven breakdown of the components of the reproduction number of online content. The Facebook Research archive (https://research.fb.com/publications) has a host of other papers further examining the spread of behaviour and content online.

Whitney Phillips's report *The Oxygen of Amplification: Better Practices for Reporting on Extremists* (Data & Society, 2018) provides a valuable summary of media manipulation efforts, and potential ways to overcome these. *Zucked: Waking Up to the Facebook Catastrophe* (HarperCollins, 2019) by Roger McNamee discusses the downsides of social media platforms, including more details on the work of Tristan Harris and Renée DiResta. 'Protecting elections from social media manipulation' by Sinan Aral and Dean Eckles (*Science*, 2019) has suggestions for ways to rigorously measure online manipulation and the potential implications for elections.

Chapter 6

For more on the origins and legacy of Mirai attack, see Garrett Graff's pair of articles for *Wired*: 'How a Dorm Room Minecraft Scam Brought Down the Internet' (2017) and 'The Mirai Botnet Architects Are Now Fighting Crime With the FBI' (2018). Landmark papers such as 'Computer Viruses – Theory and Experiments' by Fred Cohen (1984) and 'How to own the Internet in Your Spare Time' by Stuart Staniford and colleagues (*Proceedings of the 11th USENIX Security Symposium*, 2002) have more technical details on the history of viruses and worms. *Linked: The New Science of Networks* by Albert-László Barabási (Perseus, 2002) describes the history of network theory, including how networks shape malware outbreaks.

Chapter 7

'Towards a genomics-informed, real-time, global pathogen surveillance system' by Jennifer Gardy and Nick Loman (*Nature Reviews Genetics*, 2018) reviews how sequencing tools can be used to diagnose and track diseases. 'Outbreak analytics: a developing data science for informing the response to emerging pathogens' (*Philosophical Transactions of the Royal Society B*, 2019) explores the uses of data science during outbreaks, as well as areas for improvement.

Anthony Tockar's original two Neustar blog posts, 'Differential Privacy: The Basics' and 'Riding with the Stars: Passenger Privacy in the NYC Taxicab Dataset', are worth reading for a more detailed description of the New York Taxi analysis and its implications (available from: https://research.neustar.biz). *Bit By Bit: Social Research in the Digital Age* by Matthew Salganik (Princeton University Press, 2018) provides a thoughtful overview of the ethical and logical issues involved in modern social behaviour research.

Chapter 8

David Sumpter's book *Outnumbered: From Facebook and Google to Fake News and Filter-bubbles* (Bloomsbury, 2018) assesses the statistical plausibility of claims about online algorithms, with a particular focus on the Cambridge Analytica scandal. *Getting to Zero: A Doctor and a Diplomat on the Ebola Frontline* by Sinead Walsh and Oliver Johnson (Zed Books, 2018) gives a first-hand account of the politics, logistics and human cost involved in responding to the West Africa Ebola epidemic.

Acknowledgements

I'd like to thank everyone who took the time to share their expertise and experience with me while researching this book: Lucy Aplin, Nim Arinaminpathy, Wendy Barclay, Barbara Casu, Nicholas Christakis, Toby Davies, Dean Eckles, Paul Fine, Jemma Geoghegan, Andy Haldane, Heidi Larson, Rosalie Liccardo Pacula, Kristian Lum, Brendan Nyhan, Andrew Odlyzko, Whitney Phillips, John Potterat, Charlie Romford, Gary Slutkin, Briony Swire-Thompson, Jamie Tehrani, Melissa Tracy, Alex Vespignani, Charlotte Watts, and Duncan Watts. Thanks also to those who helped source historical data and documents: Victoria Cranna and Alison Forsey at the LSHTM Library & Archives Service, Liina Hultgren at the Royal Institution, and Peter Vinten-Johansen at the John Snow Archive and Research Companion. If there are any errors in the final text, they are mine alone.

I've been fortunate to have had some great mentors during my career, who have encouraged me to engage with wider audiences as well as helping me develop as a researcher: Julia Gog at the University of Cambridge, Steven Riley at Imperial College London, and John Edmunds at LSHTM. Thanks also to many, many other collaborators and colleagues I've worked with and learnt from over the years. In particular, the ideas in this book have benefitted both directly and indirectly from discussions with my brilliant colleagues in the Centre for the Mathematical Modelling of Infectious Diseases at LSHTM. As any popular science writer will know, I faced the obstacle of there being far more good research out there than I could ever fit into one book. Inevitably, I had to leave out several people and projects during the writing and editing stages, and this is of course no reflection on my views about the quality of the science.

I'd also like to thank everyone involved in the writing process. My excellent editors Cecily Gayford at Profile and Fran Barrie at Wellcome Collection have provided valuable ideas and input throughout. Thanks as well to Joe Staines for his work on copyediting the finished manuscript. And to my agent Peter Tallack, for his support and advice over the past few years. I am grateful to my parents for all their comments on initial drafts, as well as to Clare Fraser, Rachel Humby, Munir Jahangir, Stephen Rice, and Graham Wheeler for giving feedback on early chapters. Finally, I would like to thank my amazing, inspiring wife Emily, who I was lucky enough to meet while writing my last book, and lucky enough to marry while writing this one.

Index